# Generalized Transmission Line Method to Study the Far-zone Radiation of Antennas under a Multilayer Structure

# Synthesis Lectures on Antennas

Editor
Constantine A. Balanis, Arizona State University

**Generalized Transmission Line Method to Study the Far-zone Radiation of Antennas under a Multilayer Structure**
Xuan Hui Wu, Ahmed A. Kishk, and Allen W. Glisson
2008

**Narrowband Direction of Arrival Estimation for Antenna Arrays**
Jeffrey Foutz, Andreas Spanias, and Mahesh K. Banavar
2008

**Multiantenna Systems for MIMO Communications**
Franco De Flaviis, Lluis Jofre, Jordi Romeu, and Alfred Grau
2008

**Reconfigurable Antennas**
Jennifer T. Bernhard
2007

**Electronically Scanned Arrays**
Robert J. Mailloux
2007

**Introduction to Smart Antennas**
Constantine A. Balanis and Panayiotis I. Ioannides
2007

**Antennas with Non-Foster Matching Networks**
James T. Aberle and Robert Loepsinger-Romak
2007

**Implanted Antennas in Medical Wireless Communications**
Yahya Rahmat-Samii and Jaehoon Kim
2006

Generalized Transmission Line Method to Study the Far-zone Radiation of Antennas under a Multilayer Structure

Xuan Hui Wu, Ahmed A. Kishk, and Allen W. Glisson

ISBN: 978-3-031-00410-0   paperback
ISBN: 978-3-031-01538-0   ebook

DOI 10.1007/978-3-031-01538-0

A Publication in the Springer series
*SYNTHESIS LECTURES ON ANTENNAS*

Lecture #9
Series Editor: Constantine A. Balanis, Arizona State University

Series ISSN
Synthesis Lectures on Antennas
ISSN pending.      Print 1930-0328   Electronic 1930-0336

# Generalized Transmission Line Method to Study the Far-zone Radiation of Antennas under a Multilayer Structure

Xuan Hui Wu, Ahmed A. Kishk, and Allen W. Glisson

Department of Electrical Engineering
University of Mississippi

*SYNTHESIS LECTURES ON ANTENNAS #9*

# ABSTRACT

This book gives a step-by-step presentation of a generalized transmission line method to study the far-zone radiation of antennas under a multilayer structure. Normally, a radiation problem requires a fullwave analysis which may be time consuming. The beauty of the generalized transmission line method is that it transforms the radiation problem for a specific type of structure, say the multilayer structure excited by an antenna, into a circuit problem that can be efficiently analyzed. Using the Reciprocity Theorem and far-field approximation, the method computes the far-zone radiation due to a Hertzian dipole within a multilayer structure by solving an equivalent transmission line circuit. Since an antenna can be modeled as a set of Hertzian dipoles, the method could be used to predict the far-zone radiation of an antenna under a multilayer structure. The analytical expression for the far-zone field is derived for a structure with or without a polarizer. The procedure of obtaining the Hertzian dipole model that is required by the generalized transmission line method is also described. Several examples are given to demonstrate the capabilities, accuracy, and efficiency of this method.

# KEYWORDS

Transmission Line, Antenna, High Directivity, Multilayer Structure, Optimization, EBG, Hertzian Dipole, Reciprocity, Asymptotic Boundary Conditions, Far-field Radiation, Polarizer.

# Contents

# CHAPTER 1

# Introduction

Highly directive antennas are required for many wireless communication and radar applications. Normally, people adopt a horn antenna, antenna array, reflector antenna, or reflectorarray antenna to achieve the high directivity. Another easy way to increase the antenna directivity is to introduce a multilayer dielectric structure as a superstrate. This kind of structure is widely used in microwave and millimeter wave devices for fabrication convenience provided by the technologies such as multilayer printed circuit board (PCB) and low-temperature co-fired ceramic (LTCC). For example, a typical 1-D Electromagnetic Bandgap (EBG) structure is made of multilayer dielectric slabs [1]. Proper arrangement of the slabs gives transmission of the electromagnetic waves in a narrow beamwidth normal to the multilayer structure, thus achieving a high directivity. Several types of antennas were used to excite the multilayer structure, such as a patch antenna [2, 3], dipole antenna [4], dual polarized patch antenna array [5], and circularly polarized slot antenna array [6].

Most studies on the antennas embedded in a multilayer structure use fullwave analysis such as the Finite Difference Time Domain (FDTD), Method of Moments (MOM), Finite Element Method (FEM), or Modal Analysis, to simulate the entire structure including the antenna and the multilayer structure. This is very time-consuming, and therefore makes the design procedure cumbersome. In 1987, a transmission line method was devised to compute the far-field radiation of a horizontal electric Hertzian dipole embedded in a multilayer dielectric slab, and was used to investigate the gain enhancement of printed circuit antenna [7]. This method calculates the far-zone electric field by analyzing an equivalent transmission line circuit of the multilayer structure without using fullwave analysis. Recently, the transmission line method is generalized to deal with a set of arbitrarily directed electric or magnetic Hertzian dipoles embedded in a multilayer structure with or without polarizers [8, 9]. Because an antenna could be modeled by a set of Hertzian dipoles, this method is advantageous in efficiently calculating the far-field radiation of an arbitrary antenna embedded in a multilayer structure. This book presents the transmission line method in a systematic way for the study of antennas embedded in a multilayer structure. It is organized as follows.

Chapter 2 presents the transmission line method to investigate the far-field radiation of an antenna under a multilayer dielectric slab. The Reciprocity Theorem and far-field approximation are applied to transform the evaluation of the far-zone field due to the near-zone source into the evaluation of the near-zone field due to a far-zone source. Instead of a time-consuming fullwave analysis, the problem is analyzed by solving a transmission line circuit. This method is derived analytically, and verified using the commercial software package IE3D [10]. As examples, dielectric superstrates above a monopole antenna and a dielectric resonant antenna are optimized using this method to achieve high directivity.

Chapter 3 generalizes the transmission line method to make it applicable for a structure with a polarized multilayer structure. Although the derivation is for one specific type of polarizer made of perfectly conducting electric (PEC) strips with both width and period electrically thin, it is pointed out that the method can easily be extended for other types of polarizers. In contrast to a structure without a polarizer as described in Chapter 2, the plane waves in the TE and TM modes are coupled in this case so that each PEC strip interface is modeled as a four-port network, while each dielectric layer is modeled as two two-port networks, one for the TE mode and one for the TM mode. As examples, one layer of PEC strips is used to eliminate the cross polarization component of a dielectric resonator antenna (DRA) and four layers of PEC strips are used to rotate the antenna polarization. The ideas of the cross polarization reduction and rotation using PEC strip interfaces are verified using the commercial software package HFSS [11].

Chapter 4 describes the procedure to obtain a set of Hertzian dipoles that model the radiation characteristics of an antenna and are required by the generalized transmission line method. The procedures to get a narrowband model and a wideband model are given separately. Each dipole is defined by seven parameters consisting of the dipole location $(x', y', z')$, dipole direction $(\theta', \phi')$, and dipole moment $\chi' \exp(j\psi')$. For the narrowband model, these dipole parameters are constants independent of frequency, and are obtained using an optimization technique by letting the near field data due the dipole model approach to those due to the real antenna. Three such narrowband dipole models are used to replace the excitation antenna in the transmission line method in Chapters 2 and 3. For the wideband dipole model, each dipole parameter is a polynomial in the free space phase constant, and the polynomial coefficients are obtained using an optimization technique. In general, for a structure with known Green's function, either numerical or analytical, the dipole model can be used to efficiently predict the field distribution due to an antenna within the structure. As another example, a wideband Hertzian dipole model is obtained for a stacked DRA above an infinite ground plane and is then successfully used to predict the radiation patterns of the same DRA above a finite ground plane.

CHAPTER 2

# Antennas Under a Multilayer Dielectric Slab

## 2.1 INTRODUCTION

The transmission line (TL) method is presented in this chapter to study the far-zone radiation of an arbitrary Hertzian dipole embedded in a multilayer dielectric structure. It applies the Reciprocity Theorem and far-field approximation to transform the radiation problem into a plane wave scattering problem that can be solved by analyzing an equivalent transmission line circuit. This method could be used to investigate a multilayer structure excited by a physical antenna if only the antenna can be replaced by a set of Hertzian dipoles. Assuming the dielectric structure has little disturbance of the mode excited within the physical antenna, these dipoles can be obtained by sampling the current on the antenna or using an optimization technique to recover the near field surrounding the antenna after removing the dielectric structure. Thus, once the equivalent Hertzian dipoles are obtained to represent a physical antenna, the effect of different multilayer structures on the far-field radiation can be efficiently studied by using the TL method.

This chapter is organized as follows. Section 2.2 gives the derivation of the TL method for the far-field radiation due to an electric Hertzian dipole embedded in a multilayer structure. Section 2.3 presents the derivation for a magnetic Hertzian dipole. In Section 2.4, the TL method is verified using the commercial software package IE3D, where an electric Hertzian dipole is modeled by an electrically small dipole, and a magnetic Hertzian dipole modeled by an electrically small loop. Section 2.5 provides two applications of the TL method. First, the thickness and position of a dielectric slab above a quarter wavelength thin wire monopole antenna are optimized to get a high directivity. The Hertzian dipoles representing the monopole are obtained by sampling the current along the wire. After that, the arrangement of a four-layer structure is determined to maximize the directivity of a DRA. The equivalent Hertzian dipoles for the DRA are obtained by means of an optimization technique. Section 2.6 is the conclusion.

## 2.2 RADIATION DUE TO AN ELECTRIC DIPOLE

The original problem of interest is illustrated in Figure 2.1(a). An electric Hertzian dipole ($\vec{J}$) is embedded into the $k$th layer of an infinite dielectric slab with $N$ homogeneous layers. The dipole is located at the coordinate of $(x_J, y_J, z_J)$, and directed in the $(\phi_J, \theta_J)$ direction. Each dielectric layer has the thickness of $d_m$, and is characterized by relative permeability of $\mu_{rm}$ and relative permittivity of $\epsilon_{rm}$, where $m \in [1, N]$ is the layer number. The multi-layered slab is backed with an infinite plate characterized by surface impedance of $Z_L^{TE}$ for a TE mode plane wave, and $Z_L^{TM}$ for a TM mode plane wave. A special case is $Z_L^{TE} = Z_L^{TM} = 0\,\Omega$ if the backing plate is a perfect electric conductor (PEC). The far-field radiation patterns of this structure are to be computed.

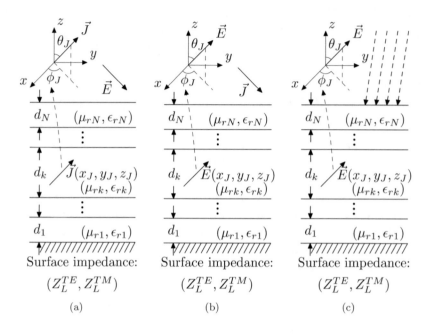

**Figure 2.1:** Problem transformation. (a) original problem, (b) transformed problem, (c) problem to be solved; from [8], copyright © 2006 IEEE.

The original radiation problem can be transformed into a scattering problem with a two-step procedure. First, the positions as well as directions of $\vec{J}$ and $\vec{E}$ in Figure 2.1(a) are exchanged, resulting in a new problem as shown in Figure 2.1(b). Based on the Reciprocity Theorem, enforcing the values of $\vec{J}$ in Figures 2.1(a) and 2.1(b) equal to each other, the values of $\vec{E}$ in the two figures are also the same. Therefore, the evaluation of $\vec{E}$ in free space due to a $\vec{J}$ in the slab is transformed into the evaluation of $\vec{E}$ in the slab due to a $\vec{J}$ in free space. Note that the $\vec{E}$ in Figure 2.1(b) is not total field, but the component in the $(\phi_J, \theta_J)$ direction. Second, the problem in Figure 2.1(b) can be further transformed into a scattering problem in Figure 2.1(c). It is done with the assumption that the $\vec{J}$ in Figure 2.1(b) is far from the slab, and the incident wave impinging on the slab can be assumed as a uniform plane wave. This assumption is valid in our case because only the far-field radiation is to be calculated for the original problem. The plane wave in Figure 2.1(c) is a TM wave if $E_\theta$ is to be evaluated in Figure 2.1(a), and a TE wave if $E_\phi$ is to be evaluated.

The $\vec{E}$ in Figure 2.1(c) can be obtained by projecting the total field at $(x_J, y_J, z_J)$ onto the $(\phi_J, \theta_J)$ direction. The total field is decomposed into a horizontal component ($\vec{E}_h$) parallel to the

$x$-$y$ plane, and a vertical component ($\vec{E}_v$) in the $z$ direction. The horizontal component has been successfully calculated in [7] by mapping the scattering problem into a transmission line circuit that is analyzed by a chain matrix method. That method is extended in this chapter to get the vertical component. The method proposed in [7] is briefly reviewed next, not only to make the discussion complete, but also because of the fact that the evaluation of the vertical component is based on the evaluation of the horizontal component. Then, the procedure to obtain the vertical component, and the projection of the total field onto the ($\phi_J, \theta_J$) direction are presented. After that, an alternative analysis adopting the S chain matrix method is proposed, also for the computation of the horizontal component. The S chain matrix method offers some advantages to the chain matrix analysis.

### 2.2.1   EVALUATION OF THE HORIZONTAL COMPONENT USING CHAIN MATRIX

The mapping from the multi-layered slab to a transmission line circuit is illustrated in Figure 2.2, where each dielectric layer is mapped into a transmission line segment (TLS) whose length equals to the slab thickness. The length, phase constant, and characteristic impedance of the TLS are represented by $d_m$, $\beta_m$, and $R_{cm}$, respectively, where $m \in [1, N]$ is the TLS number. The circuit is excited by a voltage source $V_s$ with an internal impedance $R_s$, and loaded with an impedance $Z_L$ that is equal to the surface impedance of the backing plate. $I_N$ and $I_0$ are the current flowing through the voltage source and load, respectively. From a microwave circuit point of view, the horizontal component $E_h$ in the slab equals to the $V_x$ in the circuit, where $E_h$ and $V_x$ are observed at the same altitude as shown in Figure 2.2. Therefore, the evaluation of $E_h$ in the slab is equivalent to the evaluation of $V_x$ in the circuit.

All the quantities $V_s$, $R_s$, $R_{ck}$, and $Z_L$ are dependent on the mode (TE or TM) of the exciting plane wave, as indicated in Figure 2.2, while $\beta_k$ is mode independent because the propagation directions of the plane wave in the TE and TM modes are the same, both being determined by Snell's law. Given ($\phi_i, \theta_i$) as the incident angle of the exciting plane wave, the circuit parameters are obtained as below:

$$V_s^{TE} = -j\omega\mu_o \exp(-jk_o R)/(2\pi R) \tag{2.1}$$

$$V_s^{TM} = V_s^{TE} \cos\theta_i \tag{2.2}$$

$$R_s^{TE} = \sqrt{\mu_o/\epsilon_o}/\cos\theta_i \tag{2.3}$$

$$R_s^{TM} = \cos\theta_i \sqrt{\mu_o/\epsilon_o} \tag{2.4}$$

$$R_{ck}^{TE} = \frac{\mu_{rk}\sqrt{\mu_o/\epsilon_o}}{\sqrt{\mu_{rk}\epsilon_{rk} - \sin^2\theta_i}} \tag{2.5}$$

$$R_{ck}^{TM} = \frac{\sqrt{\mu_{rk}\epsilon_{rk} - \sin^2\theta_i}}{\epsilon_{rk}\sqrt{\epsilon_o/\mu_o}} \tag{2.6}$$

$$\beta_k = k_o\sqrt{\mu_{rk}\epsilon_{rk} - \sin^2\theta_i} \ . \tag{2.7}$$

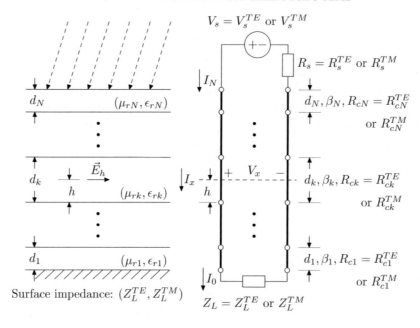

**Figure 2.2:** Transmission line circuit to calculate the horizontal $\vec{E}$ due to plane wave incidence; from [8], copyright © 2006 IEEE.

All of them are derived from the horizontal components of the electric and magnetic fields for a traveling wave in the corresponding dielectric layer. Particularly, $V_s = 2E_h^+$ in free space, and $R_{ck} = E_h^+/H_h^+$ in the $k$th dielectric layer, where $E_h^+$ and $H_h^+$ are the horizontal components of the electric and magnetic fields for the forward plane wave.

In accordance to Figure 2.2, given $A$, $B$, $C$, and $D$ as the chain matrix elements for all the cascaded TLSs, and $A'$, $B'$, $C'$, and $D'$ the chain matrix elements for the TLSs below the dashed line, two sets of linear equations are obtained as

$$\begin{bmatrix} V_s - I_N R_s \\ I_N \end{bmatrix} = \begin{bmatrix} A & B \\ C & D \end{bmatrix} \begin{bmatrix} I_0 Z_L \\ I_0 \end{bmatrix} \tag{2.8}$$

$$\begin{bmatrix} V_x \\ I_x \end{bmatrix} = \begin{bmatrix} A' & B' \\ C' & D' \end{bmatrix} \begin{bmatrix} I_0 Z_L \\ I_0 \end{bmatrix} . \tag{2.9}$$

By solving for $I_0$ in (2.8), and substituting it into (2.9), $V_x$ and $I_x$ can be obtained as

$$V_x = V_s(A'Z_L + B')/(CZ_L R_s + DR_s + AZ_L + B) \tag{2.10}$$

$$I_x = V_s(C'Z_L + D')/(CZ_L R_s + DR_s + AZ_L + B) . \tag{2.11}$$

An important point that was not mentioned in [7] is that an additional phase term should be added onto the phase of (2.10) in order to get $E_h$, especially when there are multiple Hertzian dipoles radiating. Taking the origin $(0, 0, 0)$ of the coordinate system as a reference, the phase delay due to the vertical displacement of the $\vec{J}$ in Figure 2.1(a) has already been included by analyzing the transmission line circuit. However, the phase delay due to the horizontal displacement is not included, and should be added. Thus, an exponential term is multiplied to (2.10), in order to get the phase correction for $E_h$:

$$E_h = V_x \exp[jk_o \sin\theta_i (x_J \cos\phi_i + y_J \sin\phi_i)] . \tag{2.12}$$

The polarization of $\vec{E}_h$ is in the direction of $\hat{\phi}_i$ for the TE mode excitation, and $\hat{\phi}_i \times \hat{z}$ for the TM mode excitation. This information is required when projecting the total electric field onto the $(\phi_J, \theta_J)$ direction.

## 2.2.2 EVALUATION OF THE VERTICAL COMPONENT

The vertical electric field $\vec{E}_v$ is nonzero only for the TM mode plane wave excitation, and its derivation is carried out by separating the plane wave in the $k$th dielectric layer into forward and backward waves, as shown in Figure 2.3, where $\theta_k$ is the reflected angle in that layer. The forward wave propagates from the ceiling of the layer to the floor, with a horizontal component $E_h^+$ and a vertical component $E_v^+$. The backward wave propagates from the floor to the ceiling, with a horizontal component $E_h^-$ and a vertical component $E_v^-$. Based on the fact that the vector summation of $E_h^+$ and $E_v^+$ is perpendicular to the propagation direction of the forward wave, $E_v^+$ can be obtained as

$$E_v^+ = E_h^+ \tan\theta_k = E_h^+ \frac{\sin\theta_i}{\sqrt{\epsilon_{rk}\mu_{rk} - \sin^2\theta_i}} . \tag{2.13}$$

Similarly, $E_v^-$ can be obtained as

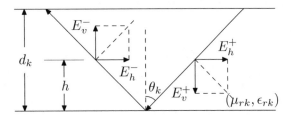

**Figure 2.3:** Wave propagation in the slab, TM mode; from [8], copyright © 2006 IEEE.

$$E_v^- = E_h^- \tan\theta_k = E_h^- \frac{\sin\theta_i}{\sqrt{\epsilon_{rk}\mu_{rk} - \sin^2\theta_i}} . \tag{2.14}$$

The total vertical component $\vec{E}_v$ is then obtained as

$$\vec{E}_v = \hat{z}(E_v^- - E_v^+) = \hat{z}(E_h^- - E_h^+)\frac{\sin\theta_i}{\sqrt{\epsilon_{rk}\mu_{rk} - \sin^2\theta_i}} . \tag{2.15}$$

In order to get $\vec{E}_v$ in (2.15), $E_h^+$ and $E_h^-$ should be separated, and this can be done by separating $V_x^+$ and $V_x^-$ because they are related with the phase correction term as

$$\begin{cases} E_h^+ = V_x^+ \exp[jk_o \sin\theta_i(x_J \cos\phi_i + y_J \sin\phi_i)] \\ E_h^- = V_x^- \exp[jk_o \sin\theta_i(x_J \cos\phi_i + y_J \sin\phi_i)] \end{cases} . \tag{2.16}$$

The voltage separation required for (2.16) can be done using either (2.17) or (2.18) below:

$$\begin{cases} V_x^+ + V_x^- = V_x \\ V_x^-/V_x^+ = \Gamma \end{cases} \implies \begin{cases} V_x^+ = V_x/(1+\Gamma) \\ V_x^- = \Gamma V_x/(1+\Gamma) \end{cases} \tag{2.17}$$

$$\begin{cases} I_x^+ - I_x^- = I_x \\ I_x^-/I_x^+ = \Gamma \end{cases} \implies \begin{cases} I_x^+ = I_x/(1-\Gamma) \\ I_x^- = \Gamma I_x/(1-\Gamma) \end{cases} \implies \begin{cases} V_x^+ = R_{ck}I_x/(1-\Gamma) \\ V_x^- = R_{ck}\Gamma I_x/(1-\Gamma), \end{cases} \tag{2.18}$$

where $V_x$ as well as $I_x$ are already determined through (2.10) and (2.11), respectively, and $\Gamma$ is the reflection coefficient calculated as $\Gamma = (Z_x - R_{ck})/(Z_x + R_{ck})$, where $Z_x = V_x/I_x$. However, problems arise in evaluating both (2.17) and (2.18), for some values of $\Gamma$. For example, if $V_x$ is evaluated at a position of short circuit so that $\Gamma = -1$, (2.17) becomes a 0/0 form that is indeterminate. A similar phenomenon occurs in (2.18) when $I_x$ is evaluated at a position of open circuit. An appropriate choice of (2.17) and (2.18) solves this problem: (2.17) is adopted for $\Gamma$ lying on the right part of the complex plane, (2.18) is adopted for $\Gamma$ lying on the left part of the complex plane, and for $\Gamma$ with pure imaginary values, either one is applicable.

## 2.2.3   FIELD PROJECTION

The total field with two components of $E_h$ and $E_v$ is projected onto the $(\phi_J, \theta_J)$ direction, to get the $\vec{E}$ in Figure 2.1. This is done in (2.19) below for the incident plane wave in the TE mode, and in (2.20) for the wave in the TM mode. As mentioned before, $E^{TE}$ and $E^{TM}$ are equal to $E_\phi$ and $E_\theta$, respectively, in the original problem in Figure 2.1(a).

$$E^{TE} = E_\phi = E_h\hat{\phi}_i \cdot (\hat{x}\cos\phi_J + \hat{y}\sin\phi_J)\sin\theta_J = E_h\sin\theta_J\sin(\phi_J - \phi_i)$$
$$\text{where } \hat{\phi}_i = -\hat{x}\sin\phi_i + \hat{y}\cos\phi_i \tag{2.19}$$

$$E^{TM} = E_\theta = (E_h\hat{\phi}_i \times \hat{z} + E_v\hat{z}) \cdot (\hat{x}\sin\theta_J\cos\phi_J + \hat{y}\sin\theta_J\sin\phi_J + \hat{z}\cos\theta_J)$$
$$= E_h\sin\theta_J\cos(\phi_J - \phi_i) + E_v\cos\theta_J . \tag{2.20}$$

### 2.2.4    EVALUATION OF THE HORIZONTAL COMPONENT USING S CHAIN MATRIX

The S chain matrix can be used to analyze the circuit in Figure 2.2 as an alternative method. Instead of calculating the total voltage, this analysis directly calculates the forward and backward voltage, thus making (2.17) and (2.18) redundant. Before adopting the S chain matrix, a requirement for the selection of nominal impedance is discussed first.

Two directly connected networks are shown in Figure 2.4, where $(a_2, b_2)$ and $(a_1', b_1')$ are the scattering parameters at the output port of network 1, and at the input port of network 2, respectively. $a_2$ and $b_2$ are obtained with a nominal impedance $R_1$, while $a_1'$ and $b_1'$ are obtained with impedance $R_2$. Because the two ports are directly connected, the voltage and current on the ports should be continuous, so

$$\begin{cases} (a_2 + b_2)\sqrt{R_1} = (a_1' + b_1')\sqrt{R_2} \\ (a_2 - b_2)/\sqrt{R_1} = (b_1' - a_1')/\sqrt{R_2} \end{cases}. \tag{2.21}$$

In addition, two conditions

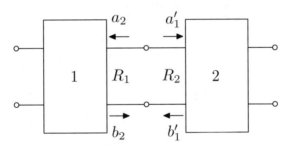

**Figure 2.4:** Two connected networks; from [8], copyright © 2006 IEEE.

$$a_2 = b_1' \text{ and } b_2 = a_1' \tag{2.22}$$

should be satisfied in order to make use of the S chain matrix. Plugging (2.22) into (2.21) gives

$$R_1 = R_2. \tag{2.23}$$

Thus, the requirement for the use of the S chain matrix is that **the nominal impedance for the connected ports should be the same.** For the circuit in Figure 2.2, where multiple TLSs with different characteristic impedances are directly connected, there are two ways to select the nominal impedance, both satisfying (2.23).

The first possibility is to choose the nominal impedance to be the same value, for example 1 Ω, for all the TLSs. Then, the TLSs can be directly connected as done in Figure 2.4. With this unit nominal impedance, the S chain matrix for the $k$th TLS is given by

$$\begin{bmatrix} a_1 \\ b_1 \end{bmatrix} = \frac{1}{4R_{ck}} \cdot \begin{bmatrix} GR_{ck} + W(1 + R_{ck}^2) & W(1 - R_{ck}^2) \\ W(R_{ck}^2 - 1) & GR_{ck} - W(1 + R_{ck}^2) \end{bmatrix} \begin{bmatrix} b_2 \\ a_2 \end{bmatrix}, \tag{2.24}$$

where $G = 4\cos(\beta_k d_k)$ and $W = j2\sin(\beta_k d_k)$. The derivation of (2.24) can be found in Appendix A.1.

The second possibility is to choose the TLS characteristic impedance as the nominal impedance. It simplifies the S chain matrix of the $k$th TLS to

$$\begin{bmatrix} a_1 \\ b_1 \end{bmatrix} = \begin{bmatrix} \exp(j\beta_k d_k) & 0 \\ 0 & \exp(-j\beta_k d_k) \end{bmatrix} \begin{bmatrix} b_2 \\ a_2 \end{bmatrix}. \tag{2.25}$$

However, in this case, the TLSs cannot be directly connected as in Figure 2.4 due to the violation of (2.23). Instead, an imaginary network ($T$) should be inserted between the adjacent TLSs, as shown in Figure 2.5, where the nominal impedance for the input port of $T$ matches that for the output port of the network 1, while the nominal impedance for the output port of $T$ matches that for the input port of the network 2. Thus, (2.23) is satisfied. The S chain matrix of $T$ is given as

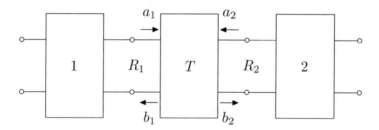

**Figure 2.5:** Impedance matching for networks with different characteristics impedance; from [8], copyright © 2006 IEEE.

$$\begin{bmatrix} a_1 \\ b_1 \end{bmatrix} = \frac{1}{2} \begin{bmatrix} \sqrt{\dfrac{R_2}{R_1}} + \sqrt{\dfrac{R_1}{R_2}} & \sqrt{\dfrac{R_2}{R_1}} - \sqrt{\dfrac{R_1}{R_2}} \\ \sqrt{\dfrac{R_2}{R_1}} - \sqrt{\dfrac{R_1}{R_2}} & \sqrt{\dfrac{R_2}{R_1}} + \sqrt{\dfrac{R_1}{R_2}} \end{bmatrix} \begin{bmatrix} b_2 \\ a_2 \end{bmatrix}. \tag{2.26}$$

It is obtained by enforcing the equality of both voltage and current between its input and output ports as

$$\begin{cases} (a_1 + b_1)\sqrt{R_1} = (a_2 + b_2)\sqrt{R_2} \\ (a_1 - b_1)/\sqrt{R_1} = (b_2 - a_2)/\sqrt{R_2} \end{cases} . \tag{2.27}$$

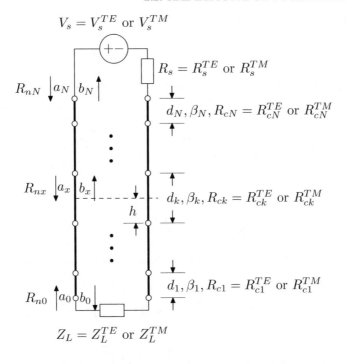

**Figure 2.6:** Transmission line circuit for the S chain matrix analysis; from [8], copyright © 2006 IEEE.

For interpretation convenience, the transmission line circuit is labeled with scattering parameters, and redrawn in Figure 2.6. $a_N$ and $b_N$ are the scattering parameters at the input port of the entire circuit, obtained with the nominal impedance $R_{nN}$, while $a_0$ and $b_0$ are for the output port of the entire circuit, with the nominal impedance $R_{n0}$. $a_x$ and $b_x$ are the scattering parameters at the input port of the circuit below the dashed line, obtained with the nominal impedance $R_{nx}$. With (2.24) or the combination of (2.25) and (2.26), the S chain matrix of several connected TLSs can be easily obtained by matrix multiplication. In accordance with Figure 2.6, given $P$, $Q$, $U$, and $V$ as the S chain matrix elements for all the cascaded TLSs, and $P'$, $Q'$, $U'$, as well as $V'$ as the S chain matrix elements for the TLSs below the dashed line, two sets of linear equations are obtained as

$$\begin{bmatrix} a_N \\ b_N \end{bmatrix} = \begin{bmatrix} P & Q \\ U & V \end{bmatrix} \begin{bmatrix} b_0 \\ a_0 \end{bmatrix} \text{ and } \begin{bmatrix} a_x \\ b_x \end{bmatrix} = \begin{bmatrix} P' & Q' \\ U' & V' \end{bmatrix} \begin{bmatrix} b_0 \\ a_0 \end{bmatrix}. \tag{2.28}$$

In (2.28), $a_0$ can be expressed in terms of $b_0$ as

$$a_0 = \Gamma_L b_0, \text{ where } \Gamma_L = (Z_L - R_{n0})/(Z_L + R_{n0}), \tag{2.29}$$

and $a_N$ expressed in terms of $b_N$ as

$$a_N = \left[ V_s \sqrt{R_{nN}} + (R_s - R_{nN})b_N \right]/(R_s + R_{nN}) . \tag{2.30}$$

The derivations of (2.29) and (2.30) can be found in Appendix A.2 and A.3, respectively. Consequently, by eliminating $a_0$ and $a_N$ with (2.29) and (2.30), (2.28) can be rewritten as

$$\begin{bmatrix} \dfrac{V_s \sqrt{R_{nN}} + (R_s - R_{nN})b_N}{R_s + R_{nN}} \\ b_N \end{bmatrix} = \begin{bmatrix} P & Q \\ U & V \end{bmatrix} \begin{bmatrix} b_0 \\ b_0 \Gamma_L \end{bmatrix} \tag{2.31}$$

$$\begin{bmatrix} a_x \\ b_x \end{bmatrix} = \begin{bmatrix} P' & Q' \\ U' & V' \end{bmatrix} \begin{bmatrix} b_0 \\ b_0 \Gamma_L \end{bmatrix} . \tag{2.32}$$

By solving (2.31) for $b_0$, and substituting the result into (2.32), $a_x$ and $b_x$ are obtained as

$$a_x = \frac{(P' + Q'\Gamma_L)V_s \sqrt{R_{nN}}}{(P + Q\Gamma_L)(R_s + R_{nN}) - (U + V\Gamma_L)(R_s - R_{nN})} \tag{2.33}$$

$$b_x = \frac{(U' + V'\Gamma_L)V_s \sqrt{R_{nN}}}{(P + Q\Gamma_L)(R_s + R_{nN}) - (U + V\Gamma_L)(R_s - R_{nN})} . \tag{2.34}$$

With $a_x$ and $b_x$, $V_x^+$ and $V_x^-$ can be derived from the relations

$$\begin{cases} (a_x + b_x)\sqrt{R_{nx}} = V_x^+ + V_x^- \\ (a_x - b_x)/\sqrt{R_{nx}} = (V_x^+ - V_x^-)/R_{ck} \end{cases} \tag{2.35}$$

as

$$V_x^+ = \frac{a_x}{2}\left( \sqrt{R_{nx}} + \frac{R_{ck}}{\sqrt{R_{nx}}} \right) + \frac{b_x}{2}\left( \sqrt{R_{nx}} - \frac{R_{ck}}{\sqrt{R_{nx}}} \right) \tag{2.36}$$

$$V_x^- = \frac{a_x}{2}\left( \sqrt{R_{nx}} - \frac{R_{ck}}{\sqrt{R_{nx}}} \right) + \frac{b_x}{2}\left( \sqrt{R_{nx}} + \frac{R_{ck}}{\sqrt{R_{nx}}} \right) . \tag{2.37}$$

Then, $E_h^+$ and $E_h^-$ in (2.16) are obtained after the phase correction, and $E_h = E_h^+ + E_h^-$. $E_v$ can be calculated from (2.15).

## 2.3  RADIATION DUE TO A MAGNETIC DIPOLE

The derivation for the far-field radiation due to a magnetic Hertzian dipole is very similar to that due to an electric Hertzian dipole. Herein, the reciprocity between a magnetic dipole $\vec{M}$ and magnetic field $\vec{H}$ is used to transform the original radiation problem into a scattering problem. It is done by simply replacing the $\vec{J}$ and $\vec{E}$ of Figure 2.1 by $\vec{M}$ and $\vec{H}$, respectively. With a representation similar to $\vec{J}$, $\vec{M}$ is located at the coordinate of $(x_M, y_M, z_M)$, and directed in the $(\phi_M, \theta_M)$ direction. If $H_\theta$ is to be evaluated in the original problem, the incident plane wave in the scattering problem is in the TE mode. For the evaluation of $H_\phi$ in the original problem, the corresponding incident wave is in the TM mode. Again, similar to what is done for the evaluation of electric field, the horizontal component of the magnetic field $H_h$ in the slab can be obtained by solving the equivalent transmission line circuit, and the vertical component $H_v$ obtained by separating the forward and backward waves. The evaluation of $H_h$ in the slab is equivalent to that of $I_x$ in the circuit. After obtaining $H_h$ and $H_v$, the total field is projected onto the $(\phi_M, \theta_M)$ direction to get $H_\phi$ or $H_\theta$ for the original problem.

## 2.3.1 EVALUATION OF THE HORIZONTAL COMPONENT USING CHAIN MATRIX

The equivalent transmission line circuit is the same as that in Figure 2.2, where all the parameters are left the same except for the voltage source $V_s$ that is redefined below:

$$V_s^{TM} = \frac{-j\omega\sqrt{\epsilon_o\mu_o}}{2\pi R} \cos\theta_i \exp(-jk_o R) \tag{2.38}$$

$$V_s^{TE} = \frac{-j\omega\sqrt{\epsilon_o\mu_o}}{2\pi R} \exp(-jk_o R) . \tag{2.39}$$

The circuit is analyzed with the same procedure described in (2.8)–(2.11). With $I_x$ in (2.11), $H_h$ is finally expressed in (2.40) for phase correction

$$H_h = I_x \exp[jk_o \sin\theta_i (x_M \cos\phi_i + y_M \sin\phi_i)] . \tag{2.40}$$

## 2.3.2 EVALUATION OF THE VERTICAL COMPONENT

The vertical component of the magnetic field $H_v$ is nonzero only for the TE mode incident wave, and can be calculated by separating the forward and backward waves, as shown in Figure 2.7. $H_v$ is derived as

$$\begin{aligned}
\vec{H}_v &= -\hat{z}(H_v^- + H_v^+) \\
&= -\hat{z}(H_h^- + H_h^+) \tan\theta_k \\
&= -\hat{z}(H_h^- + H_h^+) \frac{\sin\theta_i}{\sqrt{\epsilon_{rk}\mu_{rk} - \sin^2\theta_i}}
\end{aligned} \tag{2.41}$$

and $H_h^+$ as well as $H_h^-$ are obtained with phase correction as

$$\begin{cases} H_h^+ = I_x^+ \exp[jk_o \sin\theta_i (x_M \cos\phi_i + y_M \sin\phi_i)] \\ H_h^- = I_x^- \exp[jk_o \sin\theta_i (x_M \cos\phi_i + y_M \sin\phi_i)] \end{cases} . \tag{2.42}$$

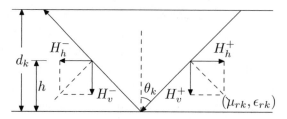

**Figure 2.7:** Wave propagation in the slab, TE mode; from [8], copyright © 2006 IEEE.

The separation of $I_x^+$ and $I_x^-$ required by (2.42) is done as in (2.43) or (2.44) below, whose derivations are very similar to those of (2.17) and (2.18):

$$\begin{cases} I_x^+ = \dfrac{V_x}{(1+\Gamma)R_{ck}} \\ I_x^- = \dfrac{\Gamma V_s}{(1+\Gamma)R_{ck}} \end{cases} \tag{2.43}$$

$$\begin{cases} I_x^+ = \dfrac{I_x}{1-\Gamma} \\ I_x^- = \dfrac{\Gamma}{1-\Gamma}I_x \end{cases}. \tag{2.44}$$

Again, the appropriate choice of (2.43) and (2.44) should be adopted based on the value of $\Gamma$ to avoid the indeterminate form of either (2.43) or (2.44) for $|\Gamma| = 1$.

### 2.3.3 FIELD PROJECTION

Similar to the electric field case, the projection of the total magnetic field onto the $(\phi_M, \theta_M)$ direction is given in (2.45) below for the TM mode wave excitation, and in (2.46) for the TE mode wave excitation. $H^{TM}$ and $H^{TE}$ are equal to $H_\phi$ and $H_\theta$, respectively, of the original problem:

$$\begin{aligned} H^{TM} = H_\phi &= H_h \hat{\phi}_i \cdot (\hat{x} \cos \phi_M + \hat{y} \sin \phi_M) \sin \theta_M \\ &= H_h \sin \theta_M \sin(\phi_M - \phi_i) \end{aligned} \tag{2.45}$$

$$\begin{aligned} H^{TE} = H_\theta &= (H_h \hat{\phi}_i \times \hat{z} + H_v \hat{z}) \cdot (\hat{x} \sin \theta_M \cos \phi_M + \hat{y} \sin \theta_M \sin \phi_M + \hat{z} \cos \theta_M) \\ &= H_h \sin \theta_M \cos(\phi_M - \phi_i) + H_v \cos \theta_M . \end{aligned} \tag{2.46}$$

### 2.3.4 EVALUATION OF THE HORIZONTAL COMPONENT USING S CHAIN MATRIX

With $V_s$ defined in (2.38) or (2.39), the S chain matrix analysis described in (2.21)–(2.37) can be used to obtain $V_x^+$ and $V_x^-$, and consequently, $I_x^+$ and $I_x^-$ as

$$I_x^+ = V_x^+ / R_{ck} \tag{2.47}$$

$$I_x^- = V_x^- / R_{ck} .$$  (2.48)

Subsequently, $H_h = H_h^+ - H_h^-$, and $H_v$ can be obtained from (2.42) and (2.41), respectively.

A convenient alternate method to solve this problem is simply to apply the Duality Theorem. The expression of $\vec{H}$ due to $\vec{M}$ is the same as that of $\vec{E}$ due to $\vec{J}$ after replacing $\epsilon$, $\mu$, $\vec{J}$ and $Z_L$ by $\mu$, $\epsilon$, $\vec{M}$ and $1/Z_L$, respectively.

## 2.4    RESULTS VERIFICATION

The far-field radiation patterns for Hertzian dipoles embedded in a three-layer dielectric structure are investigated by using the TL method, and verified using a commercial MOM software package, IE3D. The three-layer structure is infinitely extended in the $x$-$y$ plane and is backed with a PEC ground plane, as shown in Figure 2.8. The dielectric constants and the thickness of the dielectric

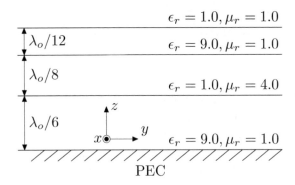

**Figure 2.8:** Three-layer dielectric structure; from [8], copyright © 2006 IEEE.

slabs are labeled in the figure, where $\lambda_o$ is the wavelength in free space. The thickness of the bottom layer is selected to be a half wavelength in the corresponding medium, while those of the top two layers are selected to be a quarter wavelength in the corresponding medium. The Hertzian dipoles exciting the multilayer structure are selected from Table 2.1, where both $J$ and $M$ are included, directed either vertically or horizontally. The subscripts of $J$ and $M$ in the first column indicate the dipole location, $o$ for a location on the $z$ axis, and $s$ for a location shifted $\lambda_o$ from the $z$ axis. The superscripts of $J$ and $M$ indicate the dipole directions. From the table, all the vertical $J$ and horizontal $M$ sources are located on the ground plane, while the horizontal $J$ and vertical $M$ sources are located a quarter wavelength above the ground plane. In order to make the radiation due to $J$ and $M$ comparable, the moment of $J$ is selected to be 1 A·m, and that of $M$ to be $\eta_0$ V·m, where $\eta_0$ is the characteristic impedance of free space.

The Hertzian dipoles in Table 2.1 are naturally accommodated in the TL method, while in IE3D, a physical structure must be used instead to emulate the Hertzian dipoles. If $\lambda_g$ is the

| Table 2.1: Hertzian dipoles; from [8], copyright © 2006 IEEE. | | | |
|---|---|---|---|
| Symbol | Type | Position $(x, y, z)$ | Moment |
| $J_o^x$ | Electric | $\left(0, 0, \dfrac{\lambda_o}{12}\right)$ | 1 A·m |
| $J_o^z$ | Electric | $(0, 0, 0)$ | 1 A·m |
| $J_s^x$ | Electric | $\left(\lambda_o \cos(\dfrac{\pi}{3}), \lambda_o \sin(\dfrac{\pi}{3}), \dfrac{\lambda_o}{12}\right)$ | 1 A·m |
| $J_s^z$ | Electric | $\left(\lambda_o \cos(\dfrac{\pi}{3}), \lambda_o \sin(\dfrac{\pi}{3}), 0\right)$ | 1 A·m |
| $M_o^x$ | Magnetic | $(0, 0, 0)$ | $\eta_0$ V·m |
| $M_o^z$ | Magnetic | $\left(0, 0, \dfrac{\lambda_o}{12}\right)$ | $\eta_0$ V·m |
| $M_s^x$ | Magnetic | $\left(\lambda_o \cos(\dfrac{\pi}{3}), \lambda_o \sin(\dfrac{\pi}{3}), 0\right)$ | $\eta_0$ V·m |
| $M_s^z$ | Magnetic | $\left(\lambda_o \cos(\dfrac{\pi}{3}), \lambda_o \sin(\dfrac{\pi}{3}), \dfrac{\lambda_o}{12}\right)$ | $\eta_0$ V·m |

wavelength in the bottom layer of the dielectric slab, a small PEC strip with length $0.02\lambda_g$ and width $0.001\lambda_g$ is used to model the electric Hertzian dipole, as shown in Figure 2.9. The dipole is excited by a voltage source located at the center of the strip. The magnetic Hertzian dipole is modeled by a square loop as illustrated in Figure 2.9, where each edge is a PEC strip with length $0.01\lambda_g$ and width $0.001\lambda_g$. In order to keep the current symmetric, each side of the loop is excited by a centered voltage source with the same internal impedance and voltage.

The far-field radiation patterns due to an electric Hertzian dipole, $J_o^x$ or $J_o^z$, are obtained by using both the TL method and IE3D, and are plotted in Figure 2.10 for the $\phi = 0°$ and $\phi = 90°$ planes. The patterns due to two electric Hertzian dipoles, $J_o^x$ & $J_s^x$, $J_o^x$ & $J_s^z$, or $J_o^z$ & $J_s^z$, are plotted in Figure 2.11 for the $\phi = 0°$ plane, and Figure 2.12 for the $\phi = 90°$ plane. Similarly, the patterns due to magnetic Hertzian dipoles are also examined. The patterns due to a magnetic Hertzian dipole, $M_o^x$ or $M_o^z$, are plotted in Figure 2.13. In addition, the patterns due to two magnetic Hertzian dipoles, $M_o^x$ & $M_s^x$, $M_o^x$ & $M_s^z$, or $M_o^z$ & $M_s^z$ are plotted in Figures 2.14 and 2.15, for the $\phi = 0°$ and $\phi = 90°$ planes, respectively.

Furthermore, the patterns due to the combination of an electric and a magnetic Hertzian dipole are also investigated. For the TL method, the analysis procedure is the same as for the previous cases. However, for IE3D, one cannot simply simulate the dipole and loop in Figure 2.9 together because the exact relationship between the dipole moments and excitation strengths is unknown for the structure. Thus, an alternative procedure is adopted. The electric and magnetic Hertzian dipoles are first simulated by IE3D separately, to get the directivity as well as the phase of far-zone field, and these results are then combined together to obtain the total radiation pattern. This procedure is described in detail as follows. The $\theta$ component of the far-zone electric field due to the combination

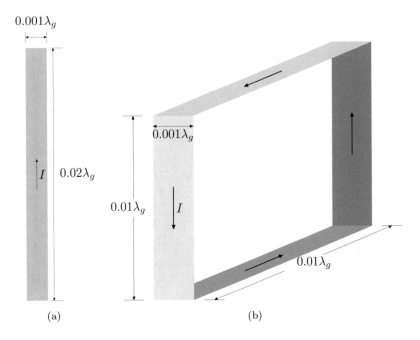

**Figure 2.9:** Hertzian dipoles in IE3D. (a) electric dipole, (b) magnetic dipole; from [8], copyright © 2006 IEEE.

of $J$ and $M$ can be obtained as

$$E^\theta_{JM}(\phi,\theta) = \sqrt{\frac{\eta_0 P_J}{4\pi R^2}} D^\theta_J(\phi,\theta) \exp[j\Phi^\theta_J(\phi,\theta)] + \sqrt{\frac{\eta_0 P_M}{4\pi R^2}} D^\theta_M(\phi,\theta) \exp[j\Phi^\theta_M(\phi,\theta)], \quad (2.49)$$

where $P_{J,M}$, $D^\theta_{J,M}$, and $\Phi^\theta_{J,M}$ are the total radiated power, partial directivity, and phase of far-zone field for the $\theta$ component, respectively. $R$ is the radius of a sphere where the electric field is evaluated. The subscript $J$ or $M$ implies a variable for an electric or a magnetic Hertzian dipole. Similarly, the $\phi$ component of the far-zone electric field due to the combination of $J$ and $M$ is obtained as

$$E^\phi_{JM}(\phi,\theta) = \sqrt{\frac{\eta_0 P_J}{4\pi R^2}} D^\phi_J(\phi,\theta) \exp[j\Phi^\phi_J(\phi,\theta)] + \sqrt{\frac{\eta_0 P_M}{4\pi R^2}} D^\phi_M(\phi,\theta) \exp[j\Phi^\phi_M(\phi,\theta)], \quad (2.50)$$

where the partial directivity $(D^\phi_{J,M})$ and phase $(\Phi^\phi_{J,M})$ are for the $\phi$ component. All the variables in (2.49) and (2.50), except for $P_J$ and $P_M$, can be obtained by using IE3D. $P_J$ and $P_M$ can be calculated as

$$P_{J,M} = \frac{1}{\eta_0} \int_0^\pi \int_0^{2\pi} \left[ |E^\theta_{J,M}(\phi,\theta)|^2 + |E^\phi_{J,M}(\phi,\theta)|^2 \right] \cdot R^2 \sin\theta \, d\phi \, d\theta, \quad (2.51)$$

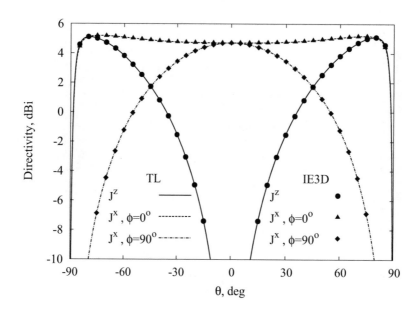

**Figure 2.10:** Radiation patterns of an electric Hertzian dipole.

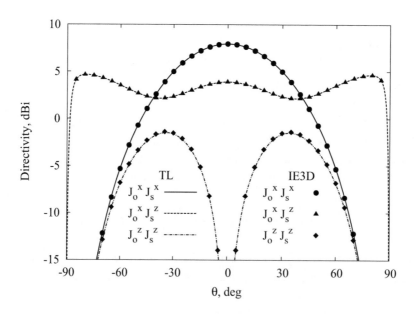

**Figure 2.11:** Radiation patterns of two electric Hertzian dipoles, $\phi = 0°$.

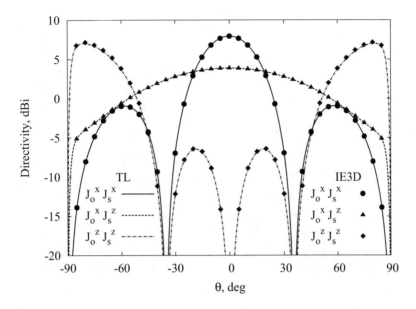

**Figure 2.12:** Radiation patterns of two electric Hertzian dipoles, $\phi = 90°$; from [8], copyright © 2006 IEEE.

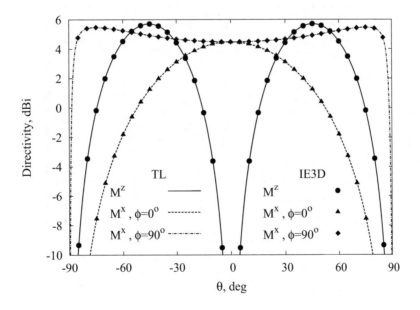

**Figure 2.13:** Radiation patterns of a magnetic Hertzian dipole.

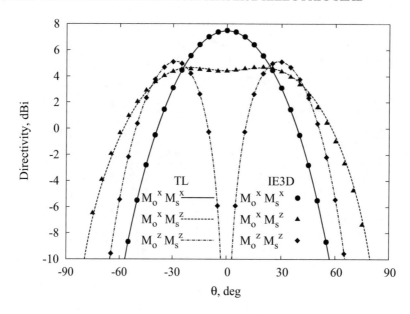

**Figure 2.14:** Radiation patterns of two magnetic Hertzian dipoles, $\phi = 0°$.

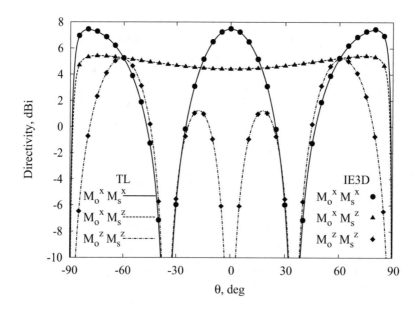

**Figure 2.15:** Radiation patterns of two magnetic Hertzian dipoles, $\phi = 90°$; from [8], copyright © 2006 IEEE.

where $E^{\theta}_{J,M}$ and $E^{\phi}_{J,M}$ are the $\theta$ and $\phi$ components, respectively, of the far-zone electric field, either due to an electric or a magnetic Hertzian dipole. Because the TL method has already been verified for the cases of a single Hertzian dipole, it can be used to evaluate (2.51). Thus, $E^{\theta}_{JM}(\phi, \theta)$ and $E^{\phi}_{JM}(\phi, \theta)$ can be determined with (2.49) and (2.50), respectively. Subsequently, the total directivity due to the combination of $J$ and $M$ can be obtained as

$$D_{JM}(\phi, \theta) = \frac{\left[|E^{\theta}_{JM}(\phi, \theta)|^2 + |E^{\phi}_{JM}(\phi, \theta)|^2\right] 4\pi R^2}{\eta_0 P_{JM}}, \tag{2.52}$$

where $P_{JM} \approx P_J + P_M$. The total radiated power ($P_{JM}$) is approximated by $P_J + P_M$. This approximation is valid for the dipoles separated by a large distance, and only affects the accuracy of the level, but not the shape, of the radiation pattern. The patterns due to the combination of an electric and a magnetic Hertzian dipole, $M^x_o$ & $J^x_s$, $M^x_o$ & $J^z_s$, $M^z_o$ & $J^x_s$, and $M^z_o$ & $J^z_s$, are calculated using the TL method, and are verified using IE3D followed by the procedure described in (2.49)–(2.52). The results are plotted in Figures 2.16 and 2.17 for the $\phi = 0°$ and $\phi = 90°$ planes, respectively. It can be seen that all the results shown in Figures 2.12–2.17 demonstrate virtually perfect agreement between the TL method and IE3D.

## 2.5    APPLICATIONS

Generally, antennas can be modeled by a set of Hertzian dipoles. Thus, the TL method could be used to investigate the far-field radiation of an arbitrary antenna embedded in a multilayer dielectric structure. The Hertzian dipoles used to represent a physical antenna will be different for different antennas, and for different multilayer structures due to the mutual coupling between the antenna and the multilayer structure. Therefore, in order to obtain the Hertzian dipoles that represent the antenna, a full wave analysis is required to simulate the entire structure including both the antenna and the multilayer structure. However, in our study, such coupling is assumed to affect only the matching but not current distribution of the physical antenna. This approximation provides a way to avoid the full wave analysis of the entire structure after changing the multilayer structure. It will be seen later that this approximation provides good accuracy in calculating the radiation pattern. On the other hand, the input impedance or matching of an antenna within a multilayer structure still requires a full wave analysis.

With the above approximation in mind, the calculation of the radiation patterns for an arbitrary antenna embedded in a multilayer dielectric structure has three steps. First, the antenna alone is analyzed by a full wave analysis, to obtain the current distribution on, or the near-fields surrounding the antenna. Second, a set of Hertzian dipoles representing the physical antenna is obtained by simply sampling the current distribution, or using some optimization technique as in [12]. Last, the set of dipoles is inserted into the multilayer dielectric structure to replace the physical antenna, and the TL method is used to calculate the radiation pattern. In the remaining part of this section, a quarter wavelength thin wire monopole antenna and a DRA, both embedded in a multilayer dielectric structure, are studied through this three-step procedure.

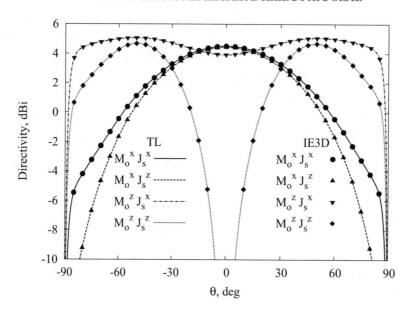

**Figure 2.16:** Radiation patterns of an electric and a magnetic Hertzian dipoles, $\phi = 0°$.

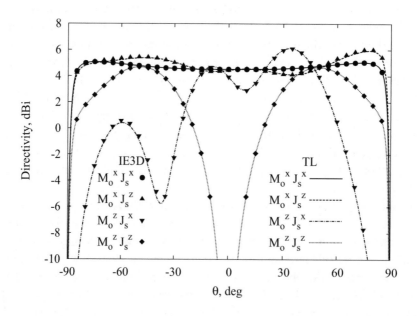

**Figure 2.17:** Radiation patterns due to an electric and a magnetic Hertzian dipole, $\phi = 90°$; from [8], copyright © 2006 IEEE.

## 2.5.1    THIN WIRE MONOPOLE ANTENNA IN A TWO-LAYER STRUCTURE

Consider a quarter wavelength thin wire monopole antenna embedded into a two-layer structure as shown in Figure 2.18, where the two-layer structure is backed with the PEC ground plane. The bottom layer is filled with air and the top layer is filled with dielectric material of $\epsilon_r = 10.2$ and $\mu_r = 1.0$. The thicknesses of the two layers are to be optimized, in order to maximize the directivity of the monopole antenna.

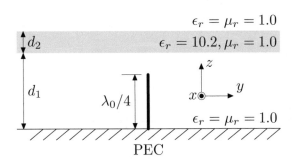

**Figure 2.18:** A quarter wavelength thin wire monopole antenna in a two-layer structure; from [8], copyright © 2006 IEEE.

Before applying the TL method, the monopole antenna should be replaced by a set of electric Hertzian dipoles, which can be obtained by sampling the current along the wire. Assuming the presence of the dielectric slab in Figure 2.18 does not affect the current distribution along the wire, the MOM is used to simulate the thin wire monopole without the presence of the dielectric slab. A total of twenty current values evenly distributed along the wire are sampled, thus providing 20 vertically directed electric Hertzian dipoles to approximate the monopole with the presence of the dielectric slab. After that, the TL method is applied to efficiently find the optimal $d_1 = 0.713\lambda_0$ and $d_2 = 0.238\lambda_g$, which leads to the maximum directivity of 9.218 dBi in the $\theta = \pm 45°$ directions. Given the optimal $d_1$ and $d_2$, the radiation pattern obtained by using the TL method is plotted in Figure 2.19 and is compared to the results obtained from a full wave analysis of the entire structure by IE3D. Perfect agreement can be observed. The analyses are performed using a Dell Precision 360 machine. IE3D requires 4.3 minutes and 10 megabytes of memory when the entire geometry is meshed with 40 cells per wavelength. In comparison, the TL method requires only 0.42 s if the field symmetry along the $\phi$ direction is used to calculate the total radiated power as a normalization factor, and 29.2 s if such symmetry is not used. The memory required for the TL method is only around 1.16 megabytes. In the TL method, a 0.5° angle increment is used to numerically calculate the total radiated power.

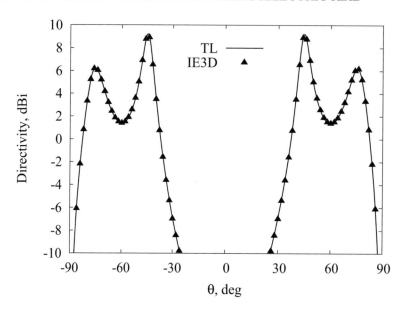

**Figure 2.19:** Radiation pattern of the quarter wavelength thin wire monopole antenna in the two-layer structure of Figure 2.18 with $d_1 = 0.713\lambda_0$ and $d_2 = 0.238\lambda_g$; from [8], copyright © 2006 IEEE.

## 2.5.2    DRA IN A FOUR-LAYER STRUCTURE

A DRA is embedded into a four-layer structure as shown in Figure 2.20. The DRA is a circular cylinder with a radius of 4 mm and a height of 2.5 mm, and is made of a material with $\epsilon_r = 10.2$ and $\mu_r = 1.0$. It is fed by a probe located on the $x$ axis and offset 3 mm from the center axis of the DRA. The probe has radius 0.3 mm, is 2 mm long, and is connected to a coaxial line with inner and outer radii of 0.3 mm and 0.6 mm, respectively. The four-layer structure is constructed by two dielectric slabs located separately above the ground plane, each with a thickness of a quarter wavelength. The slabs are made of the same material as that of the DRA. The TL method is used to determine the spacings $d_1$ and $d_3$ in Figure 2.20 of the two dielectric slabs so as to achieve a high directivity.

Similar to the thin wire monopole antenna case, the DRA should be replaced by a set of Hertzian dipoles in order to apply the TL method. Although these dipoles could be obtained by sampling the equivalent surface current on a surface enclosing the DRA, the number of the dipoles may be very large. Alternatively, the method proposed in [12] is used instead to obtain a small number of Hertzian dipoles representing the DRA. Briefly, the dipoles are obtained by using an optimization technique, such that the tangential electric field on a surface enclosing the DRA due to the Hertzian dipoles approaches to the exact field values, or the values that would be normally obtained by a full wave analysis. Based on the Uniqueness Theorem, the obtained set of Hertzian dipoles is equivalent to the DRA in the sense of generating the same fields. Similar to the procedure used in

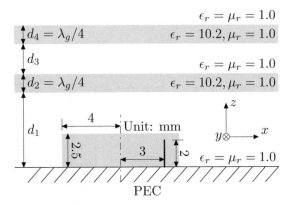

**Figure 2.20:** A DRA in a four-layer structure; from [8], copyright © 2006 IEEE.

the monopole antenna case, the two dielectric slabs are removed to obtain the Hertzian dipoles, with the assumption that the presence of the slabs does not significantly affect the current distribution. In this study, the tangential electric fields used to get the Hertzian dipoles are obtained by simulating the structure without the dielectric slabs using the WIPL-D commercial software package [13]. With the tangential electric field data, the Particle Swarm Optimization (PSO) method [14, 15], instead of the Genetic Algorithm (GA) used in [12], is adopted to find the Hertzian dipoles equivalent to the DRA.

Three different sets of Hertzian dipoles are obtained by using the PSO, each representing the DRA at the resonant frequency of 12 GHz. The sets are tabulated in Tables 2.2–2.4: Table 2.2 for the set of two electric and two magnetic Hertzian dipoles, Table 2.3 for six electric Hertzian dipoles, and Table 2.4 for five magnetic Hertzian dipoles. The optimization procedure used to get the three dipole models is presented in Chapter 4. The radiation patterns in the $\phi = 0°, 90°$ planes due to the three sets of dipoles above the ground plane are calculated, and compared to the exact patterns for the DRA on the ground plane, as shown in Figure 2.21. The "exact" patterns are obtained from WIPL-D. It can be seen that each set of the Hertzian dipoles could be used to replace the DRA in the sense of generating similar far-field radiation.

Given the Hertzian dipoles representing the DRA, the TL method can be applied to efficiently determine $d_1$ and $d_3$ in Figure 2.20 to obtain a high directivity. It is found that the directivity reaches the maximum value of 25.085 dBi for $d_1 = 0.5\lambda_0$ and $d_3 = 0.75\lambda_0$, and a second maximum value of 24.768 dBi for $d_1 = 0.5\lambda_0$ and $d_3 = 0.25\lambda_0$, both at the $\theta = 0°$ direction. Clearly, the later arrangement gives a much compact structure without losing high directivity. For $d_1 = 0.5\lambda_0$ and $d_3 = 0.25\lambda_0$, the radiation patterns in the $\phi = 0°, 90°$ planes are calculated by using the TL method, and plotted in Figures 2.22–2.24 for three sets of Hertzian dipoles. The results are also verified by IE3D, which is used to simulate the entire structure of Figure 2.20 including the four-layer structure and the real DRA. All the curves show good agreement between the TL method and IE3D.

Table 2.2: Modeling of DRA by two $J$ and two $M$; from [8], copyright © 2006 IEEE.

| Type | $x_{J,M}$, m | $y_{J,M}$, m | $z_{J,M}$, m | $\theta_{J,M}$, rad | $\phi_{J,M}$, rad | Moment |
|---|---|---|---|---|---|---|
| $J$ | 4.76448e-3 | 2.86291e-4 | 1.14876e-3 | 2.39596 | 0.171734 | $0.206736 \exp(j2.23059)$ A·m |
| $J$ | -4.53493e-3 | 1.49248e-3 | 9.55276e-5 | 1.28169 | 4.99056 | $0.034039 \exp(j0.965524)$ A·m |
| $M$ | 1.04837e-4 | -6.63911e-5 | 2.20923e-4 | 1.41244 | 4.70504 | $229.296 \exp(j0.11707)$ V·m |
| $M$ | -2.43446e-3 | 2.47014e-4 | 2.39667e-3 | 1.4651 | 1.62738 | $37.9225 \exp(j1.64629)$ V·m |

Table 2.3: Modeling of DRA by six $J$; from [8], copyright © 2006 IEEE.

| $x_J$, m | $y_J$, m | $z_J$, m | $\theta_J$, rad | $\phi_J$, rad | Moment, A·m |
|---|---|---|---|---|---|
| -2.75797e-3 | -3.14113e-3 | 9.08772e-4 | 0.486058 | 5.79757 | $0.0333375 \exp(j1.5269)$ |
| -5.50492e-4 | 5.83374e-4 | 1.03684e-3 | 0.691911 | 0.161498 | $0.734674 \exp(j1.66594)$ |
| 2.21735e-3 | 7.65622e-4 | 9.39355e-4 | 2.28962 | 0.301145 | $0.64832 \exp(j1.93143)$ |
| -8.9681e-5 | 2.07743e-3 | 1.59864e-4 | 2.05102 | 0.300332 | $0.480824 \exp(j2.03781)$ |
| 1.34076e-3 | -2.45966e-3 | 7.4472e-4 | 1.89703 | 5.64953 | $0.417378 \exp(j1.129)$ |
| 1.14854e-3 | 3.27729e-3 | 3.55705e-4 | 1.31092 | 4.05865 | $0.57721 \exp(j2.74747)$ |

Table 2.4: Modeling of DRA by five $M$; from [8], copyright © 2006 IEEE.

| $x_M$, m | $y_M$, m | $z_M$, m | $\theta_M$, rad | $\phi_M$, rad | Moment, V·m |
|---|---|---|---|---|---|
| 5.04283e-4 | 2.54641e-3 | 2.56915e-3 | 0.945148 | 4.82985 | $144.471 \exp(j0.12597)$ |
| 2.76138e-3 | 2.2041e-3 | 3.72696e-4 | 2.52507 | 3.12697 | $117.604 \exp(j0.797935)$ |
| 4.41552e-4 | -2.03298e-3 | 2.88045e-3 | 1.4394 | 1.98076 | $134.311 \exp(j3.13991)$ |
| 2.68366e-3 | -2.38996e-3 | 9.4356e-4 | 1.95805 | 6.15464 | $77.059 \exp(j0.752697)$ |
| -1.33184e-3 | -8.75822e-4 | 0 | 2.20954 | 4.36266 | $213.252 \exp(j0.00015666)$ |

On the same computer as for the last example, IE3D requires 1.37 hours and 516 megabytes of memory with a mesh scheme of 18 cells per wavelength, while the TL method requires less than 18 s and about 1.27 megabytes of memory for each dipole model. In the TL method, the angle increment for calculating the total radiated power as a normalization factor is 0.5°.

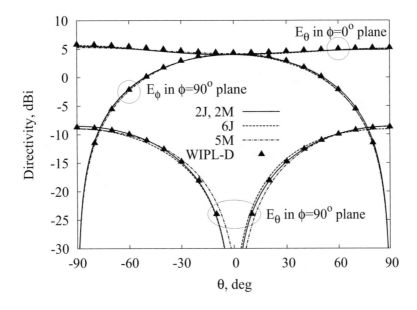

**Figure 2.21:** Radiation patterns of the DRA on the ground plane; from [8], copyright © 2006 IEEE.

## 2.6    CONCLUSIONS

A transmission line method was developed to calculate the far-field radiation of arbitrarily directed Hertzian dipoles that are embedded in a multilayer dielectric structure. Based on the Reciprocity Theorem, the evaluation of the far-field was transformed into the evaluation of the field in the multilayer structure due to an incident plane wave. The horizontal field component in the multilayer structure was derived by solving a cascaded transmission line circuit that is mapped from the multilayer structure. The analysis of the circuit was carried out by using either the chain matrix or S chain matrix. The vertical field component in the multilayer structure was fully dependent on the horizontal component, and was derived after separating the forward and backward waves inside the structure. The derivation was based on the fact that for either forward or backward wave, the vector of the total electric or magnetic field is perpendicular to the wave propagation direction. The TL method was implemented and verified by IE3D for the case of a three-layer structure excited by either electric Hertzian dipoles, magnetic Hertzian dipoles, or their combination. The TL method is advantageous in efficiently predicting the radiation patterns of any antenna embedded in a mul-

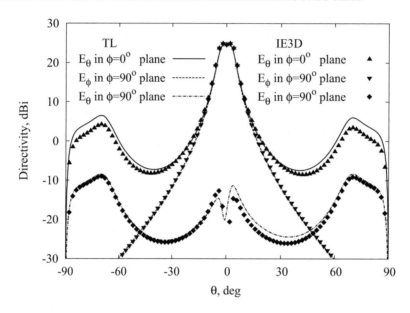

**Figure 2.22:** Radiation patterns of the DRA in a four-layer structure using the $6J$ model, $d_1 = 0.5\lambda_0$, $d_3 = 0.25\lambda_0$; from [8], copyright © 2006 IEEE.

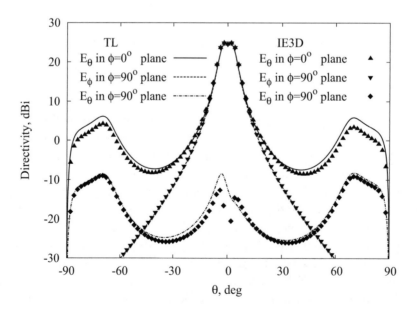

**Figure 2.23:** Radiation patterns of the DRA in a four-layer structure using the $5M$ model, $d_1 = 0.5\lambda_0$, $d_3 = 0.25\lambda_0$; from [8], copyright © 2006 IEEE.

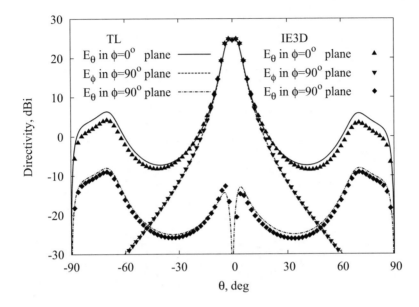

**Figure 2.24:** Radiation patterns of the DRA in a four-layer structure using the $2J2M$ model, $d_1 = 0.5\lambda_0$, $d_3 = 0.25\lambda_0$; from [8], copyright © 2006 IEEE.

tilayer dielectric structure by replacing the physical antenna with a set of Hertzian dipoles. As two application examples, the arrangement of the multilayer structure was optimized to maximize the directivity of a quarter wavelength thin wire monopole antenna and a DRA. The results were also validated by full wave analysis of the entire physical structure by using IE3D.

# CHAPTER 3

# Antennas Under a Polarized Multilayer Structure

## 3.1 INTRODUCTION

In the last chapter, a transmission line (TL) method was presented to compute the far-field radiation of arbitrary Hertzian dipoles in a multilayer dielectric structure. The TL method is based on the Reciprocity Theorem and far-field approximation to transform the evaluation of the far-field due to enforced sources in the dielectric structure into the evaluation of the field in the dielectric structure due to an incident plane wave. The transformed problem can be solved by analyzing an equivalent transmission line circuit. This method can be generalized to deal with any multilayer structure if only the structure is reciprocal and homogeneous in the plane along the structure. Specifically, an anisotropic layer that couples the TE and TM mode plane waves can be incorporated in the structure.

In this chapter, the TL method is generalized to deal with a structure embedded with periodic PEC strip interfaces. Both the strip width and period are assumed to be small compared to wavelength, and thus the strip interface can be considered as a homogeneous sheet. This kind of an interface was previously used as a soft or hard surface [16, 17]. As mentioned before, the procedure presented here is valid for any reciprocal multilayer structure if the transmission and reflection coefficients of each layer can be obtained for both the TE and TM modes. Furthermore, this method provides a way to compute the far-field radiation of a structure with both nonperiodic and periodic layers. This chapter is organized as follows. Section 3.2 presents the procedure to compute the far-field radiation due to an electric Hertzian dipole. Section 3.3 is for the case of a magnetic Hertzian dipole. Section 3.4 derives the S matrix of a strip interface. Section 3.5 gives two possible applications of the strip interface embedded structure. Section 3.6 discusses the advantages and limitations of the TL method. Section 3.7 is the conclusion.

## 3.2 RADIATION DUE TO AN ELECTRIC DIPOLE

The problem of interest is illustrated in Figure 3.1. An electric Hertzian dipole is located in a multilayer structure backed with an infinite plate characterized by a reflection matrix $\tilde{\Gamma}_L$ defined as

$$\begin{bmatrix} E_h^{TE-} & E_h^{TM-} \end{bmatrix}^T = \tilde{\Gamma}_L \begin{bmatrix} E_h^{TE+} & E_h^{TM+} \end{bmatrix}^T, \tag{3.1}$$

where $E_h^{TE+}$ and $E_h^{TM+}$ are the horizontal components of the incident electric field for the TE and TM modes, respectively, and $E_h^{TE-}$ as well as $E_h^{TM-}$ are those of the reflected electric field.

The dipole is directed in a $(\theta_J, \phi_J)$ direction, and located at a coordinate of $(x_J, y_J, z_J)$. A set of $N-1$ PEC strip interfaces (PEC-SIs) are embedded into the structure and are shown as solid bold lines in Figure 3.1. A PEC-SI is composed of periodically arranged infinitely long PEC strips. The strips are directed along the $\phi_{s,k}$ angle with the $x$ axis, where $k$ is the PEC-SI index. Both the strip width and separation are small compared to wavelength. Each broken line parallel to the PEC-SIs represents an interface between two layers of different dielectric materials. The far-field radiation due to this structure is to be computed.

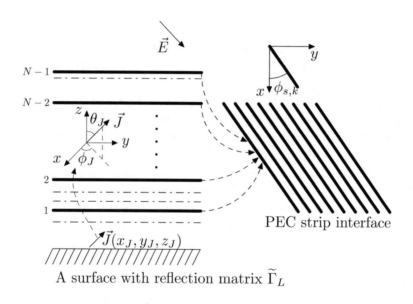

**Figure 3.1:** A multilayer structure polarized by PEC strip interfaces; from [9], copyright © 2007 IEEE.

Similar to Chapter 2, after applying the Reciprocity Theorem and far-field approximation, the evaluation of the far-zone $\vec{E}$ field due to $\vec{J}$ in the structure can be transformed into the evaluation of the $\vec{E}$ field at the source position and in the source direction due to plane wave excitations. The transformed problem can be solved by analyzing an equivalent network as shown in Figure 3.2. Each two-port network is composed of one or several cascaded transmission line segments (TLS) that are mapped from a multilayer dielectric section in Figure 3.1. Such a mapping is described in Chapter 2. The network characteristics depend on the mode of the plane wave propagating inside the dielectric slab. Specifically, given $k \in [1, N]$, $P_k^{TE}$, $Q_k^{TE}$, $U_k^{TE}$, and $V_k^{TE}$ are the S chain matrix elements of

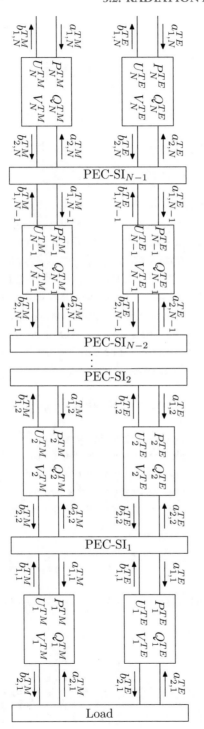

**Figure 3.2:** Equivalent network for the structure in Figure 3.1; from [9], copyright © 2007 IEEE.

the TLS circuit below the $k$th or above the $(k-1)$th PEC-SI for the TE mode, and are defined as

$$\begin{bmatrix} a_{1,k}^{TE} \\ b_{1,k}^{TE} \end{bmatrix} = \begin{bmatrix} P_k^{TE} & Q_k^{TE} \\ U_k^{TE} & V_k^{TE} \end{bmatrix} \begin{bmatrix} b_{2,k}^{TE} \\ a_{2,k}^{TE} \end{bmatrix}. \tag{3.2}$$

Similarly, $P_k^{TM}$, $Q_k^{TM}$, $U_k^{TM}$, and $V_k^{TM}$ are for the TM mode, and are defined as

$$\begin{bmatrix} a_{1,k}^{TM} \\ b_{1,k}^{TM} \end{bmatrix} = \begin{bmatrix} P_k^{TM} & Q_k^{TM} \\ U_k^{TM} & V_k^{TM} \end{bmatrix} \begin{bmatrix} b_{2,k}^{TM} \\ a_{2,k}^{TM} \end{bmatrix}. \tag{3.3}$$

The derivation of these S chain matrices can be found in Chapter 2 for lossless dielectric layers. It is also possible to incorporate lossy dielectric layers by adopting lossy transmission line segments. All of the scattering parameters are obtained with the nominal impedance equal to the characteristic impedance of the TLS port. Each four-port network in Figure 3.2 is mapped from a PEC-SI and referred to as a PEC-SI network. The S matrix of a PEC-SI network will be derived in Section 3.4. The excitation $a_{1,N}^{TE}$ and $a_{1,N}^{TM}$ are defined based on which field component is to be evaluated in Figure 3.1. If $E_\theta$ is to be computed, the plane wave impinging on the structure is due to a $\hat{\theta}$ directed $\vec{J}$ in the far-zone, and is in the TM mode. The excitation scattering parameters can be defined as $a_{1,N}^{TE} = 0$ and $a_{1,N}^{TM} = a_s$, where $a_s$ is obtained by normalizing the horizontal electric field by the square root of the nominal impedance of the input port as

$$a_s = -j\omega\mu_o \exp(-jk_o R) \cos\theta_i / (4\pi R \sqrt{\eta_o \cos\theta_i}), \tag{3.4}$$

where $R$ is the distance from the source to the observation point, $\eta_o$ is the characteristic impedance of free space, and $\theta_i$ is the radiation angle with respect to the $z$ axis. If $E_\phi$ is to be evaluated, the incident plane wave is in the TE mode, and it can be proved that $a_{1,N}^{TE} = a_s$ and $a_{1,N}^{TM} = 0$.

With the excitation and all the network characteristics defined, a set of linear equations are required to analyze the circuit in Figure 3.2. For the $k$th PEC-SI network, four linear equations can be obtained

$$\begin{bmatrix} a_{2,k+1}^{TE} & a_{2,k+1}^{TM} & a_{1,k}^{TE} & a_{1,k}^{TM} \end{bmatrix}^T = \tilde{S}_k \begin{bmatrix} b_{2,k+1}^{TE} & b_{2,k+1}^{TM} & b_{1,k}^{TE} & b_{1,k}^{TM} \end{bmatrix}^T, \tag{3.5}$$

where $\tilde{S}_k$ is the S matrix of the $k$th PEC-SI network. Plugging (3.2) and (3.3) into (3.5) eliminates $a_{1,k}^{TE}$, $b_{1,k}^{TE}$, $a_{1,k}^{TM}$, and $b_{1,k}^{TM}$:

$$\begin{bmatrix} a_{2,k+1}^{TE} \\ a_{2,k+1}^{TM} \\ a_{2,k+1}^{TM} \\ P_k^{TE} b_{2,k}^{TE} + Q_k^{TE} a_{2,k}^{TE} \\ P_k^{TM} b_{2,k}^{TM} + Q_k^{TM} a_{2,k}^{TM} \end{bmatrix} = \tilde{S}_k \begin{bmatrix} b_{2,k+1}^{TE} \\ b_{2,k+1}^{TM} \\ b_{2,k+1}^{TM} \\ U_k^{TE} b_{2,k}^{TE} + V_k^{TE} a_{2,k}^{TE} \\ U_k^{TM} b_{2,k}^{TM} + V_k^{TM} a_{2,k}^{TM} \end{bmatrix}. \tag{3.6}$$

Therefore, a total of $4(N-1)$ equations can be obtained from all the PEC-SIs. Two equations can be obtained from the load

$$
\begin{bmatrix} a_{2,1}^{TE}\sqrt{R_{c,L}^{TE}} \\ a_{2,1}^{TM}\sqrt{R_{c,L}^{TM}} \end{bmatrix} = \widetilde{\Gamma}_L \begin{bmatrix} b_{2,1}^{TE}\sqrt{R_{c,L}^{TE}} \\ b_{2,1}^{TM}\sqrt{R_{c,L}^{TM}} \end{bmatrix} ,
\tag{3.7}
$$

where $R_{c,L}^{TE}$ and $R_{c,L}^{TM}$ are the nominal impedance of the port connecting the loads for the TE and TM modes, respectively. Two additional linear equations can be obtained from the known excitation sources $a_{1,N}^{TE}$ and $a_{1,N}^{TM}$:

$$
\begin{cases} a_{1,N}^{TE} = P_N^{TE} b_{2,N}^{TE} + Q_N^{TE} a_{2,N}^{TE} \\ a_{1,N}^{TM} = P_N^{TM} b_{2,N}^{TM} + Q_N^{TM} a_{2,N}^{TM} \end{cases} .
\tag{3.8}
$$

Now, a linear system of order $4N$ can be obtained from (3.6), (3.7), and (3.8). After solving this system for the $4N$ unknowns $a_{2,k}^{TE}, b_{2,k}^{TE}, a_{2,k}^{TM}$, and $b_{2,k}^{TM}$ with $k \in [1, N]$, the voltage and current at any point in the circuit of Figure 3.2 can be determined. Then, the $E_\theta$ and $E_\phi$ in the far-zone region can be computed by projecting the total $\vec{E}$ field at the source position onto the $(\theta_J, \phi_J)$ direction following the Equations (2.12)–(2.20) in Chapter 2.

## 3.3    RADIATION DUE TO A MAGNETIC DIPOLE

Similar to Chapter 2, the radiation due to a magnetic Hertzian dipole can be obtained in two different ways. The first one is very similar to the procedure described in the previous section. After applying the reciprocity between a magnetic Hertzian dipole $\vec{M}$ and magnetic field $\vec{H}$, and the far-field approximation, the evaluation of the far-zone $\vec{H}$ field due to $\vec{M}$ in the structure can be transformed into the evaluation of the $\vec{H}$ field at the source position and in the source direction due to plane wave excitation. The horizontal component of the $\vec{H}$ field in the structure is equivalent to the current in the network in Figure 3.2. Now, the excitation scattering parameter $a_s$ that is previously defined in (3.4) is due to a magnetic Hertzian dipole in the far-zone, and should be redefined as

$$
a_s = -j\omega\sqrt{\epsilon_o\mu_o}\exp(-jk_oR)\cos\theta_i/(4\pi R\sqrt{\eta_o\cos\theta_i}) .
\tag{3.9}
$$

After applying the new excitation, the circuit in Figure 3.2 can be analyzed by building and solving a linear system from (3.6)–(3.8). The vertical component of the $\vec{H}$ field is fully dependent on the horizontal component, and can be obtained from (2.41) in Chapter 2. Then, the $H_\theta$ and $H_\phi$ in the far-zone region can be computed by projecting the total $\vec{H}$ field at the source position onto the source direction. The second way to obtain the radiation due to a magnetic Hertzian dipole is based on the Duality Theorem. However, the procedure of applying the Duality Theorem to get $\vec{H}$ due to $\vec{M}$ is different to that in Chapter 2. For a structure with PEC-SIs as shown in Figure 3.1, the expression of $\vec{H}$ due to $\vec{M}$ in the structure of Figure 3.1 is the same as that of $\vec{E}$ due to $\vec{J}$ in a dual structure after replacing $\epsilon$, $\mu$ and $\vec{J}$ by $\mu$, $\epsilon$ and $\vec{M}$, respectively. The dual structure is obtained by replacing the PEC and PMC in the original structure in Figure 3.1 by PMC and PEC, respectively. The equivalent network of the dual structure is similar to that in Figure 3.2 except that the four-port networks are for PMC strip interfaces and the reflection matrix $\widetilde{\Gamma}_L$ of the load may also be different.

## 3.4 ASYMPTOTIC BOUNDARY CONDITIONS

### 3.4.1 PEC-TYPE ASYMPTOTIC BOUNDARY CONDITIONS

A PEC-SI located between two different materials with parameters of $(\mu_{r,1}, \epsilon_{r,1})$ and $(\mu_{r,2}, \epsilon_{r,2})$ is shown on the left side of Figure 3.3. The strips are in the $\hat{p}$ direction with an angle $\phi_s$ with respect to the $x$ axis. Since the PEC-SI is anisotropic, for either a TE or TM mode plane wave impinging on it, the reflected and transmitted waves could be a combination of plane waves in TE and TM modes. Therefore, unlike a dielectric layer, which can be modeled as a two-port network either for TE or TM mode, the PEC-SI has to be modeled as a four-port network as shown on the right side of Figure 3.3. Port 1 and Port 2 are for the plane wave above the interface in the TE and TM modes, respectively. Port 3 and Port 4 are for the plane wave below the interface for the two modes. The characteristic impedances of the ports are equal to those of the equivalent transmission lines for the correspondence material and are given as $R_{c,1}^{TE} = \eta_1/\cos\theta_1$, $R_{c,2}^{TE} = \eta_2/\cos\theta_2$, $R_{c,1}^{TM} = \eta_1\cos\theta_1$, and $R_{c,2}^{TM} = \eta_2\cos\theta_2$ where $\eta_1 = \eta_o\sqrt{\mu_{r,1}/\epsilon_{r,1}}$ and $\eta_2 = \eta_o\sqrt{\mu_{r,2}/\epsilon_{r,2}}$. The S parameters of the four-port network can be obtained by exciting each port separately and evaluating the outgoing waves at all the four ports. For example, a plane wave of either TE or TM mode propagating downwards as shown in Figure 3.3 corresponds to the excitation of either Port 1 or Port 2, respectively. Since the strip width and periodicity are small compared to wavelength, the PEC-SI can be considered as a homogeneous interface and the asymptotic boundary conditions [18] are applied:

$$\begin{cases} \hat{p} \cdot (\vec{E}_i + \vec{E}_r) = 0 \\ \hat{p} \cdot \vec{E}_t = 0 \\ \hat{o} \cdot (\vec{E}_i + \vec{E}_r - \vec{E}_t) = 0 \\ \hat{p} \cdot \left( \dfrac{\hat{k}_{i,2} \times \vec{E}_i}{\eta_2} + \dfrac{\hat{k}_{r,2} \times \vec{E}_r}{\eta_2} - \dfrac{\hat{k}_{t,1} \times \vec{E}_t}{\eta_1} \right) = 0 \end{cases} , \tag{3.10}$$

where $\hat{o} = \hat{z} \times \hat{p}$ is a unit vector orthogonal to the strip and tangential to the interface. Two additional equations can be obtained based on the fact that the electric field vector of a plane wave is orthogonal to the propagation direction

$$\vec{E}_r \cdot \hat{k}_{r,2} = 0 \text{ and } \vec{E}_t \cdot \hat{k}_{t,1} = 0 . \tag{3.11}$$

Due to the phase matching for the tangential electric and magnetic fields across the PEC-SI, Snell's law still holds and can be used to determine $\hat{k}_{r,2}$ and $\hat{k}_{t,1}$ from $\hat{k}_{i,2}$. Thus, given the excitation $\vec{E}_i$ and the propagation vector of the incident plane wave $\hat{k}_{i,2}$, $\vec{E}_r$ and $\vec{E}_t$, each with three unknown scalar components, can be computed from the six equations in (3.10) and (3.11).

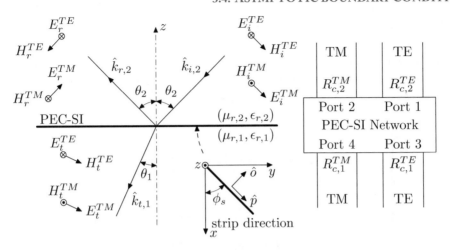

**Figure 3.3:** Equivalent network of a PEC-SI; from [9], copyright © 2007 IEEE.

Suppose the incident plane has an angle $\phi$ with the $x$-$z$ plane, for a TE plane wave excitation $\vec{E}_i = \hat{\phi} = -\hat{x}\sin\phi + \hat{y}\cos\phi$, the equivalent incident voltage at Port 1 can be obtained as $V_1^+ = \vec{E}_i \cdot \hat{\phi}$. After $\vec{E}_r$ and $\vec{E}_t$ are computed from (3.10) and (3.11) by a computer algebra system called Maxima [19], the outgoing voltage at all the ports can be obtained as

$$\begin{cases} V_1^- = \vec{E}_r \cdot \hat{\phi}, & V_2^- = \vec{E}_r \cdot (\hat{\phi} \times \hat{z}) \\ V_3^- = \vec{E}_t \cdot \hat{\phi}, & V_4^- = \vec{E}_t \cdot (\hat{\phi} \times \hat{z}) \end{cases}. \tag{3.12}$$

The Maxima code to derive the reflected and transmitted waves for the PEC-SI case is given in Appendix B.1. Consequently, the first column of the S matrix can be written as

$$\begin{cases} S_{11} = \dfrac{V_1^-}{V_1^+} = \dfrac{1}{G}\{\,[2 - (\cos\phi_d + 1)\sin^2\theta_1]\eta_2\cos\theta_2 + \\ \qquad\quad [(\cos\phi_d + 1)\sin^2\theta_2 - 2\cos\phi_d]\eta_1\cos\theta_1\} \\[2mm] S_{21} = \dfrac{V_2^-}{V_1^+}\sqrt{\dfrac{R_{c,2}^{TE}}{R_{c,2}^{TM}}} = \dfrac{2\eta_1\sin\phi_d\cos\theta_1\cos\theta_2}{G} \\[2mm] S_{31} = \dfrac{V_3^-}{V_1^+}\sqrt{\dfrac{R_{c,2}^{TE}}{R_{c,1}^{TE}}} = \dfrac{-2}{G}\sqrt{\eta_1\eta_2\cos^3\theta_1\cos^3\theta_2}\cdot(\cos\phi_d + 1) \\[2mm] S_{41} = \dfrac{V_4^-}{V_1^+}\sqrt{\dfrac{R_{c,2}^{TE}}{R_{c,1}^{TM}}} = \dfrac{2\sin\phi_d}{G}\sqrt{\eta_1\eta_2\cos\theta_1\cos^3\theta_2} \end{cases}, \tag{3.13}$$

where   $G = [(\cos\phi_d + 1)\sin^2\theta_2 - 2]\eta_1\cos\theta_1 + [(\cos\phi_d + 1)\sin^2\theta_1 - 2]\eta_2\cos\theta_2$   and   $\phi_d = 2(\phi_s - \phi)$.

For the excitation of Port 2, $\vec{E}_i = \hat{k}_{i,2} \times \hat{\phi}$ and the equivalent incident voltage $V_2^+ = \vec{E}_i \cdot (\hat{\phi} \times \hat{z})$. Similar to the procedure described above, after getting $\vec{E}_r$ and $\vec{E}_t$ from (3.10) and (3.11), and therefore the outgoing voltages from (3.12), the second column of the S matrix can be found as

$$
\begin{cases}
S_{12} = \dfrac{V_1^-}{V_2^+}\sqrt{\dfrac{R_{c,2}^{TM}}{R_{c,2}^{TE}}} = \dfrac{2\eta_1\sin\phi_d\cos\theta_1\cos\theta_2}{G} \\[4mm]
S_{22} = \dfrac{V_2^-}{V_2^+} = \dfrac{-1}{G}\{\,[(\cos\phi_d + 1)\sin^2\theta_1 - 2]\eta_2\cos\theta_2 + \\[2mm]
\qquad\qquad [(\cos\phi_d + 1)\sin^2\theta_2 - 2\cos\phi_d]\eta_1\cos\theta_1\} \\[4mm]
S_{32} = \dfrac{V_3^-}{V_2^+}\sqrt{\dfrac{R_{c,2}^{TM}}{R_{c,1}^{TE}}} = \dfrac{2\sin\phi_d}{G}\sqrt{\eta_1\eta_2\cos^3\theta_1\cos\theta_2} \\[4mm]
S_{42} = \dfrac{V_4^-}{V_2^+}\sqrt{\dfrac{R_{c,2}^{TM}}{R_{c,1}^{TM}}} = \dfrac{2}{G}\sqrt{\eta_1\eta_2\cos\theta_1\cos\theta_2}\cdot(\cos\phi_d - 1)
\end{cases} \tag{3.14}
$$

In order to get the third and fourth columns of the S matrix, the excitation plane wave of either TE or TM mode in Figure 3.3 should propagate upwards. It can be proved that the expressions for $S_{33}$, $S_{43}$, $S_{13}$, $S_{23}$, $S_{34}$, $S_{44}$, $S_{14}$, and $S_{24}$ are the same as the expressions for $S_{11}$, $S_{21}$, $S_{31}$, $S_{41}$, $S_{12}$, $S_{22}$, $S_{32}$, and $S_{42}$, respectively, after replacing the subscript indexes 1 and 2 by 2 and 1, respectively, for all the symbols with such an index. They are listed as follows:

$$
\begin{cases}
S_{13} = \dfrac{V_1^-}{V_3^+}\sqrt{\dfrac{R_{c,1}^{TE}}{R_{c,2}^{TE}}} = \dfrac{-2}{G}\sqrt{\eta_2\eta_1\cos^3\theta_2\cos^3\theta_1}\cdot(\cos\phi_d + 1) \\[4mm]
S_{23} = \dfrac{V_2^-}{V_3^+}\sqrt{\dfrac{R_{c,1}^{TE}}{R_{c,2}^{TM}}} = \dfrac{2\sin\phi_d}{G}\sqrt{\eta_2\eta_1\cos\theta_2\cos^3\theta_1} \\[4mm]
S_{33} = \dfrac{V_3^-}{V_3^+} = \dfrac{1}{G}\{\,[2 - (\cos\phi_d + 1)\sin^2\theta_2]\cos\theta_1\eta_1 + \\[2mm]
\qquad\qquad [(\cos\phi_d + 1)\sin^2\theta_1 - 2\cos\phi_d]\eta_2\cos\theta_2\} \\[4mm]
S_{43} = \dfrac{V_4^-}{V_3^+}\sqrt{\dfrac{R_{c,1}^{TE}}{R_{c,1}^{TM}}} = \dfrac{2\eta_2\sin\phi_d\cos\theta_2\cos\theta_1}{G}
\end{cases} \tag{3.15}
$$

$$\begin{cases} S_{14} = \dfrac{V_1^-}{V_4^+}\sqrt{\dfrac{R_{c,1}^{TM}}{R_{c,2}^{TE}}} = \dfrac{2\sin\phi_d}{G}\sqrt{\eta_2\eta_1\cos^3\theta_2\cos\theta_1} \\[2em] S_{24} = \dfrac{V_2^-}{V_4^+}\sqrt{\dfrac{R_{c,1}^{TM}}{R_{c,2}^{TM}}} = \dfrac{2}{G}\sqrt{\eta_2\eta_1\cos\theta_2\cos\theta_1}\cdot(\cos\phi_d - 1) \\[2em] S_{34} = \dfrac{V_3^-}{V_4^+}\sqrt{\dfrac{R_{c,1}^{TM}}{R_{c,1}^{TE}}} = \dfrac{2\eta_2\sin\phi_d\cos\theta_2\cos\theta_1}{G} \\[2em] S_{44} = \dfrac{V_4^-}{V_4^+} = \dfrac{-1}{G}\{\,[(\cos\phi_d + 1)\sin^2\theta_2 - 2]\eta_1\cos\theta_1 + \\[1em] \qquad\qquad [(\cos\phi_d + 1)\sin^2\theta_1 - 2\cos\phi_d]\eta_2\cos\theta_2\} \ . \end{cases} \tag{3.16}$$

It can be shown that the S matrix is symmetrical such that $S_{ij} = S_{ji}$ for $i \neq j$.

## 3.4.2    PMC-TYPE ASYMPTOTIC BOUNDARY CONDITIONS

Similar to a PEC-SI, a PMC strip interface (PMC-SI) can also be modeled as a four-port network as shown in Figure 3.3 where the PEC strips are replaced by PMC strips and the equivalent network is referred to as a PMC-SI network. This equivalent network can be used to compute the far-field radiation due to a magnetic dipole embedded in a structure with PEC-SIs based on the Duality Theorem as discussed in Section 3.3. The PMC-SI is considered as a homogeneous interface and the PMC-type asymptotic boundary conditions are applied:

$$\begin{cases} \hat{p}\cdot(\hat{k}_{i,2}\times\vec{E}_i + \hat{k}_{r,2}\times\vec{E}_r) = 0 \\ \hat{p}\cdot(\hat{k}_{t,1}\times\vec{E}_t) = 0 \\ \hat{o}\cdot\left(\dfrac{\hat{k}_i\times\vec{E}_i}{\eta_2} + \dfrac{\hat{k}_r\times\vec{E}_r}{\eta_2} - \dfrac{\hat{k}_t\times\vec{E}_t}{\eta_1}\right) = 0 \\ \hat{p}\cdot(\vec{E}_i + \vec{E}_r - \vec{E}_t) = 0 \end{cases} \ . \tag{3.17}$$

Using (3.17) and (3.11), the S parameters of the PMC-SI network can be obtained in the similar manner as that for the PEC-SI network, and are given below:

$$\begin{cases} S_{11} = \dfrac{1}{G'}\{[(\cos\phi_d + 1)\sin^2\theta_1 - 2]\eta_1\cos\theta_2 + \\[1em] \qquad\quad [(\cos\phi_d + 1)\sin^2\theta_2 - 2\cos\phi_d]\eta_2\cos\theta_1\} \\[1.5em] S_{21} = \dfrac{2\eta_2\sin\phi_d\cos\theta_1\cos\theta_2}{G'} \\[1.5em] S_{31} = \dfrac{2(\cos\phi_d - 1)}{G'}\sqrt{\eta_1\eta_2\cos\theta_1\cos\theta_2} \\[1.5em] S_{41} = \dfrac{-2\sin\phi_d}{G'}\sqrt{\eta_1\eta_2\cos^3\theta_1\cos\theta_2} \end{cases} \tag{3.18}$$

$$\begin{cases} S_{12} = \dfrac{2\eta_2 \sin\phi_d \cos\theta_1 \cos\theta_2}{G'} \\[2mm] S_{22} = \dfrac{1}{G'}\left\{[(\cos\phi_d + 1)\sin^2\theta_1 - 2]\eta_1 \cos\theta_2 + \right. \\[2mm] \qquad\quad \left. [2\cos\phi_d - (\cos\phi_d + 1)\sin^2\theta_2]\eta_2 \cos\theta_1\right\} \\[2mm] S_{32} = \dfrac{-2\sin\phi_d}{G'}\sqrt{\eta_1\eta_2 \cos\theta_1 \cos^3\theta_2} \\[2mm] S_{42} = \dfrac{-2(\cos\phi_d + 1)}{G'}\sqrt{\eta_1\eta_2 \cos^3\theta_1 \cos^3\theta_2} \end{cases} \tag{3.19}$$

$$\begin{cases} S_{13} = \dfrac{2(\cos\phi_d - 1)}{G'}\sqrt{\eta_2\eta_1 \cos\theta_2 \cos\theta_1} \\[2mm] S_{23} = \dfrac{-2\sin\phi_d}{G'}\sqrt{\eta_2\eta_1 \cos^3\theta_2 \cos\theta_1} \\[2mm] S_{33} = \dfrac{1}{G'}\left\{[(\cos\phi_d + 1)\sin^2\theta_2 - 2]\eta_2 \cos\theta_1 + \right. \\[2mm] \qquad\quad \left. [(\cos\phi_d + 1)\sin^2\theta_1 - 2\cos\phi_d]\eta_1 \cos\theta_2\right\} \\[2mm] S_{43} = \dfrac{2\eta_1 \sin\phi_d \cos\theta_2 \cos\theta_1}{G'} \end{cases} \tag{3.20}$$

$$\begin{cases} S_{14} = \dfrac{-2\sin\phi_d}{G'}\sqrt{\eta_2\eta_1 \cos\theta_2 \cos^3\theta_1} \\[2mm] S_{24} = \dfrac{-2(\cos\phi_d + 1)}{G'}\sqrt{\eta_2\eta_1 \cos^3\theta_2 \cos^3\theta_1} \\[2mm] S_{34} = \dfrac{2\eta_1 \sin\phi_d \cos\theta_2 \cos\theta_1}{G'} \\[2mm] S_{44} = \dfrac{1}{G'}\left\{[(\cos\phi_d + 1)\sin^2\theta_2 - 2]\eta_2 \cos\theta_1 + \right. \\[2mm] \qquad\quad \left. [2\cos\phi_d - (\cos\phi_d + 1)\sin^2\theta_1]\eta_1 \cos\theta_2\right\} \;, \end{cases} \tag{3.21}$$

where $G' = [(\cos\phi_d + 1)\sin^2\theta_2 - 2]\eta_2 \cos\theta_1 + [(\cos\phi_d + 1)\sin^2\theta_1 - 2]\eta_1 \cos\theta_2$ and $\phi_d$ is the same as that for (3.13). The Maxima code to derive the reflected and transmitted waves for the PMC-SI case is given in Appendix B.2.

## 3.5    APPLICATIONS

The TL method can be used to investigate the radiation characteristics of an antenna embedded in a multilayer structure with PEC strip interfaces after replacing the physical antenna by a set of Hertzian dipoles. The set of dipoles is normally obtained by using an optimization technique, and has similar radiation performance to that of the physical antenna in both the near and far regions.

### 3.5.1    CROSS POLARIZATION REDUCTION

The first application is shown in Figure 3.4, where a DRA is mounted above an infinite ground plane and a dielectric slab is put a half wavelength above the ground plane. The DRA is a circular cylinder with a radius of 4 mm and a height of 2.5 mm, and is fed by a 2 mm long probe located on the $x$ axis and 3 mm offset from the center axis. The dielectric slab has a thickness of a quarter wavelength, and is covered by a strip interface with $\phi_s = 90°$ on the bottom surface. Both the DRA and the dielectric slab are made of a material with $\epsilon_r = 10.2$ and $\mu_r = 1.0$. In Chapter 2, three different dipole models consisting of two electric and two magnetic dipoles (2J2M), six electric dipoles (6J), and five magnetic dipoles (5M) have been obtained for the DRA on the infinite ground plane at 12 GHz, and are listed in Table 2.2, Table 2.3 and Table 2.4, respectively. After replacing the DRA by any of the dipole models and applying the TL method, the radiation patterns of the structure in Figure 3.4 can be obtained, and are plotted in Figures 3.5 and 3.6 for the cases without and with the strip interface, respectively. Clearly, the dielectric slab itself increases the antenna directivity while the $y$ directed PEC strips completely eliminate the cross polarization component that is $E_\theta$ in the $\phi = 90°$ plane. Practically, due to the finite width and period of the PEC strips, the cross polarization component will not be zero, but at a very low level. The results presented here are obtained using the 2J2M model. Although not shown here to save space, the 5M and 6J models give almost the same results.

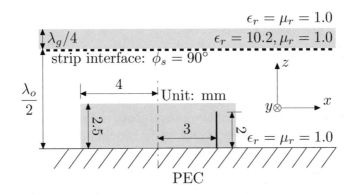

**Figure 3.4:** Cross polarization reduction by using a PEC-SI; from [9], copyright © 2007 IEEE.

For an infinite multilayer structure without PEC-SIs, the TL method was verified in Chapter 2 by using the commercial software package IE3D, which is based on the multilayer Green's function. However, we are not aware of any software package available which is able to give radiation patterns for a structure with infinite PEC strips. Although the Finite-Difference Time-Domain (FDTD) method or the Finite Element method (FEM) can be used to compute the input impedance of the structure in Figure 3.4 if the perfect matched layer (PML) is used to truncate the infinite structure, neither method is able to get the radiation pattern unless the Green's function for the PEC-SIs embedded multilayer structure is provided. Therefore, a truncated version of the structure in Figure 3.4 is shown in Figure 3.7, and is simulated by applying the FEM based commercial software package HFSS. The gain enhancement of a finite dielectric slab was studied in [20, 21]. In

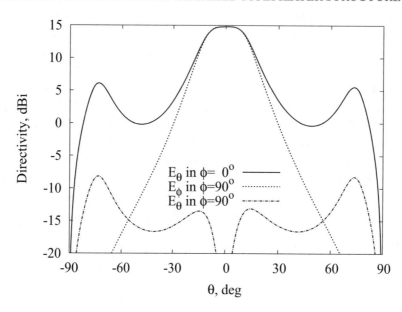

**Figure 3.5:** Radiation patterns of the structure in Figure 3.4 without the PEC-SI; from [9], copyright © 2007 IEEE.

Figure 3.7, the DRA is mounted on the infinite ground plane, and is covered by a 60 mm × 60 mm dielectric slab with $\hat{y}$ directed PEC strips printed on the bottom surface. All the other dimensions and the material parameters are the same as those in Figure 3.4. In order to reduce the leakage of cross-polarized radiation, the four side walls of the structure are also covered by PEC strips, which are $y$ directed in the two surfaces parallel to the $y$-$z$ plane and $z$ directed in the two surfaces parallel to the $x$-$z$ plane. The PEC strip has a width of 0.4 mm and a period of 0.8 mm. The radiation patterns of the structure in Figure 3.7 are obtained by using HFSS, and are plotted in Figures 3.8 and 3.9 for the cases without and with the PEC strips, respectively. It can be seen that the PEC strips eliminate the cross polarization, and the finite dielectric slab gives almost the same directivity in the $\theta = 0°$ direction as that predicted by the TL method in Figure 3.6 for an infinite structure. A smaller dielectric slab may not give the high directivity as shown in Figures 3.8 and 3.9.

## 3.5.2    POLARIZER

The second application is illustrated in Figure 3.10, where the DRA and dielectric slab are the same as those in Figure 3.4 but a different PEC-SI arrangement is chosen. Four PEC-SIs are inserted into the structure. They are closely placed and evenly separated with a 1 mm distance. The bottom PEC-SI is located at a height of $h$ above the ground plane. The strip directions of the PEC-SIs from the top to the bottom are 0°, 22.5°, 45°, and 67.5°, respectively. The 2J2M model is used to replace

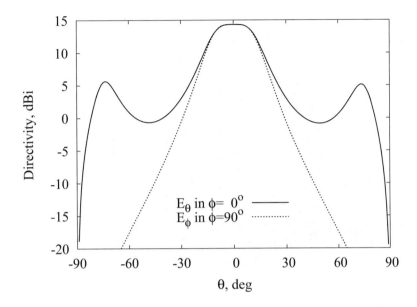

**Figure 3.6:** Radiation patterns of the structure in Figure 3.4 with the PEC-SI; from [9], copyright © 2007 IEEE.

the DRA and then the TL method is applied to efficiently analyze the structure in Figure 3.10. The radiation patterns are plotted in Figure 3.11 for $h = 7.5$ mm and Figure 3.12 for $h = 9.5$ mm. It can be seen that the set of four sequentially rotated PEC-SIs works as a polarizer that rotates the antenna polarization, and the position of the polarizer greatly affects the antenna's directivity. Again, the 5M and 6J models give the same results although they are not shown here.

As for the last application, we have no commercial software package available to compute the radiation patterns for the structure in Figure 3.10. Instead, HFSS is used to simulate a truncated structure as shown in Figure 3.13, where a 40 mm × 40 mm dielectric slab is used to cover the DRA above the infinite ground plane and the polarizer is located at a height of $h = 9.5$ mm. Similar to the last application, the side walls are covered by PEC strips to reduce the leakage of cross-polarized radiation, but the strip directions are different to those in Figure 3.7. Each strip has a width of 0.4 mm. The strip period is 0.8 mm for the top PEC-SI and the two side walls parallel to the $y$-$z$ plane, 1.0 mm for the two side walls parallel to the $x$-$z$ plane, and 1.6 mm for the other three PEC-SIs in order to reduce the memory requirement for solving the problem. The structure is simulated by using a workstation equipped with eight gigabytes of memory and two 64 bit processors, each at 3.6 GHz, and the resulting radiation patterns are plotted in Figure 3.14. It can be seen that the truncated structure realizes the polarization rotation predicted by the TL method, and gives almost the same directivity in the $\theta = 0°$ direction as that in Figure 3.12, but the cross polarization level is

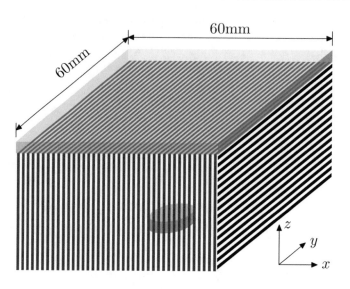

**Figure 3.7:** Structure simulated by using HFSS, for cross polarization reduction; from [9], copyright © 2007 IEEE.

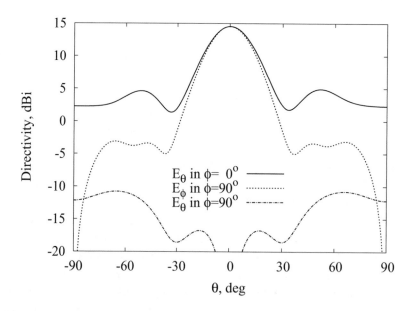

**Figure 3.8:** Radiation patterns of the structure in Figure 3.7 without the PEC strips; from [9], copyright © 2007 IEEE.

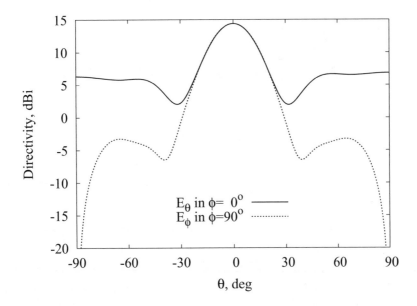

**Figure 3.9:** Radiation patterns of the structure in Figure 3.7 with the PEC strips; from [9], copyright © 2007 IEEE.

higher. Obviously, the cross polarization can be reduced by applying thinner and denser PEC strips, but that increases the size of the system matrix and cannot be solved by our available computer.

## 3.6    DISCUSSION

The TL method is not limited to the study of a multilayer structure with PEC strip interfaces. It is applicable for a structure with arbitrary homogeneous and reciprocal layers or interfaces of either isotropic or anisotropic if only the S matrices, $\tilde{S}_k$ in (3.5) and (3.6), are provided. Such a layer or interface may be a two-dimensional periodic structure with its period small compared to wavelength in both the mediums above and below the structure, so that it can be treated as a homogeneous layer or interface whose S matrix can be obtained by using a full-wave analysis equipped with a periodic boundary condition (PBC). Given a plane wave excitation impinging on the periodic layer or interface, the reflected and transmitted electric fields can be obtained numerically by simulating a unit cell of the structure using a PBC equipped full analysis instead of the analytical solution for the PEC strips in Section 3.4. Then, the outgoing voltages at the four ports can be obtained from (3.12), with which the S matrix can be easily computed. There are two types of structures that can be studied by the TL method but not by a PBC equipped full-wave analysis: a structure with several periodic interfaces which have no common period, and a structure with both periodic and nonperiodic layers as shown in Figures 3.4 and 3.10 where the layer with the DRA is nonperiodic.

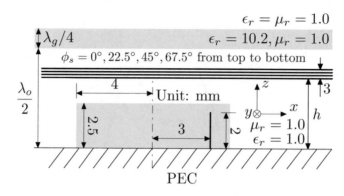

**Figure 3.10:** Polarization rotation by using four PEC-SIs; from [9], copyright © 2007 IEEE.

However, as explained in Chapter 2, the TL method only deals with the far field so far because it makes use of far-field approximation to convert the Hertzian dipole excitation into a plane wave excitation. The TL method has another limitation when it is applied to a structure with PEC-SIs. A PEC-SI has a total reflection of the incident wave if the electric field is aligned to the strip direction, and may support a cavity mode. Although the cavity mode is not excited by a plane wave outside the structure due to reciprocity, it does appear in the solution of the linear system with Equations (3.6), (3.7), and (3.8). The existence of a cavity mode makes the system matrix singular and thus the method fails. Numerically, the cavity mode is difficult to remove from the solution. On the other hand, the investigation of the system matrix singularity can be used to help us in practical design. For example, in the design of the radiating structures in Figures 3.4 and 3.10, the cavity mode should be avoided because it traps the wave inside the structure and thus reduces radiation efficiency. So, the TL method can be used to optimize a multilayer structure to avoid the cavity mode.

## 3.7   CONCLUSIONS

The TL method was generalized to compute the far-field radiation of Hertzian dipoles in a multilayer structure embedded with PEC strip interfaces. The problem was solved by analyzing an equivalent network, where each PEC strip interface was mapped into a four-port network, and each dielectric layer was mapped into two two-port networks, one for the TE mode and one for the TM mode. The S matrix of a four-port network was obtained analytically by applying the asymptotic boundary conditions assuming the strip width and period are small compared to wavelength. The network was analyzed by solving a linear system whose order depends on the number of PEC strip interfaces. As an application, a dielectric slab with PEC strips printed on its bottom surface was put above a DRA to increase the antenna directivity and eliminate cross polarization. Another application is to change the antenna polarization by using four sequentially rotated PEC strip interfaces. For each of the two applications, a truncated version of the infinite structure was simulated by using

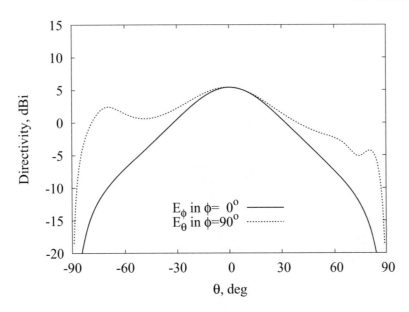

**Figure 3.11:** Radiation patterns of the structure in Figure 3.10 with $h = 7.5$ mm; from [9], copyright © 2007 IEEE.

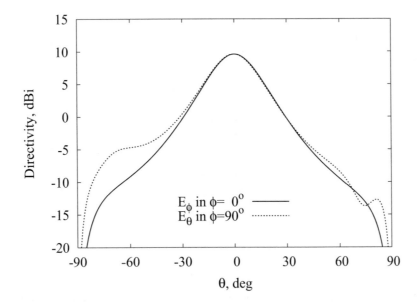

**Figure 3.12:** Radiation patterns of the structure in Figure 3.10 with $h = 9.5$ mm; from [9], copyright © 2007 IEEE.

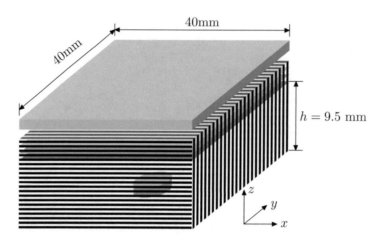

**Figure 3.13:** Structure simulated by using HFSS, for polarization rotation; from [9], copyright © 2007 IEEE.

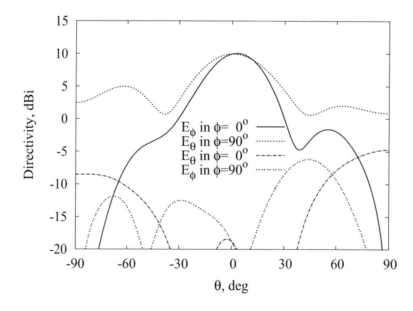

**Figure 3.14:** Radiation patterns of the structure in Figure 3.13; from [9], copyright © 2007 IEEE.

HFSS, which validated the ideas of cross-polarization reduction and polarization rotation by using PEC strips. It was pointed out that any homogeneous and reciprocal layer or interface, not only the PEC strip interface, can be inserted into a multilayer structure which can then be solved by the TL method. In addition, the TL method was able to solve two kinds of problems which cannot be done by a periodic boundary condition equipped full-wave analysis. Furthermore, although the TL method cannot remove the possible cavity mode in the solution, the investigation of the system matrix singularity helps us to avoid the cavity mode in designing a radiating structure.

# CHAPTER 4

# Hertzian Dipole Model for an Antenna

## 4.1    INTRODUCTION

An antenna is normally studied by using a full wave analysis to get the input impedance and radiation characteristics. For the convenience of system analysis, an equivalent circuit model is often extracted from the antenna input impedance, so that the system including the antenna can be studied with circuit theory assuming no coupling between the antenna and the circuit connected to it. However, such a circuit model gives no information about the antenna radiation performance. It has been found that a set of Hertzian dipoles can be obtained by using optimization techniques to recover the near field of an antenna, thus modeling its radiation characteristics [12, 22]. Such a dipole model provides a simple and convenient way to study the radiation performance of an antenna embedded in a specific structure whose Green's function is known, assuming no interaction between the antenna and the structure. In Chapters 2 and 3, three Hertzian dipole models for a dielectric resonator antenna (DRA) above an infinite ground plane have been obtained, and have been found valid for the DRA embedded in a multilayer structure. The dipole models used in Chapters 2 and 3 are valid only within a narrow frequency band because all the dipole parameters are constants independent of frequency and the near-field data at a single frequency are used in the optimization. Therefore, those models are referred to as narrowband Hertzian dipole models.

   In this chapter, the procedure of getting a narrowband Hertzian dipole model and a wideband Hertzian dipole model are proposed separately. For a narrowband dipole model, the dipole parameters are constants independent of frequency. The dipole parameters are the unknowns to be determined by minimizing the difference between the near field due to the dipoles and that due to the physical antenna. On the other hand, the procedure to get a wideband model is more complicated. For a wideband model, all the dipole parameters are frequency dependent as polynomials in the phase constant, and are able to track the variation of the near field with frequency. The polynomial coefficients are the unknowns to be determined. Different to the narrowband model and many other optimization problems, the unknown coefficients for the wideband model can not be used to construct the optimization parametric space because of lack of knowledge about their ranges, which is required by an optimization updating scheme. Instead, the parametric space is constructed by using the values of the dipole parameters at different frequencies. Given a point in such a parametric space, the polynomial coefficients for each dipole parameter can be obtained by solving a linear system. After defining the parametric space and the objective or fitness function, an optimization algorithm can be applied to search for the optimum solution. Herein, the Particle Swarm Optimization (PSO) method is adopted [14] for both the narrowband dipole model and wideband dipole model.

This chapter is organized as follows. Section 4.2 presents the procedure to get a narrowband dipole model. Section 4.3 demonstrates the limitations of the narrowband model. Section 4.4 describes the optimization procedure to obtain a wideband dipole model. A wideband stacked DRA is then modeled as an example, and an application of the wideband dipole model is given. The noise rejection capability and frequency scalability of the dipole model are demonstrated subsequently. Section 4.5 is the conclusion.

## 4.2    NARROWBAND HERTZIAN DIPOLE MODEL

Obtaining a narrowband dipole model is an optimization problem, and the PSO method is chosen here to solve it. In this section, the framework of the PSO method is presented to solve a general multivariate and uniobjective optimization problem, followed by its implementation to obtain the dipole model.

### 4.2.1    PARTICLE SWARM OPTIMIZATION METHOD

A multivariate and uniobjective optimization problem can be briefly described as follows:

Given $\alpha^{\min} = [\alpha_1^{\min}, \alpha_2^{\min}, \ldots, \alpha_n^{\min}]$ and $\alpha^{\max} = [\alpha_1^{\max}, \alpha_2^{\max}, \ldots, \alpha_n^{\max}] \in \mathcal{R}^n$ where $\alpha_i^{\min} \leq \alpha_i^{\max}$ for $i = 1, 2, \ldots, n$, a parametric space is defined as $\mathcal{W} = \{[\alpha_1, \alpha_2, \ldots, \alpha_n] | \alpha_i \in [\alpha_i^{\min}, \alpha_i^{\max}]\}$. With a fitness function $\mathcal{F}$ that projects the parametric space into the real set: $\mathcal{W} \xrightarrow{\mathcal{F}} \mathcal{R}$, the optimization objective is to find $\alpha_{\mathrm{opt}} \in \mathcal{W}$ such that $\mathcal{F}(\alpha_{\mathrm{opt}}) \leq \mathcal{F}(\alpha) \ \forall \alpha \in \mathcal{W}$.

The PSO method solves the above problem by using $m$ particles, $b_1, b_2, \ldots, b_m$, simultaneously moving in the parametric space $\mathcal{W}$ and searches for $\alpha_{\mathrm{opt}}$. For an arbitrary particle $b_k, k = 1, 2, \ldots, m$, it has the information of its position $\alpha^k = [\alpha_1^k, \alpha_2^k, \ldots, \alpha_n^k] \in \mathcal{W}$, instantaneous velocity $v^k = [v_1^k, v_2^k, \ldots, v_n^k] \in \mathcal{R}^n$, the best solution it has found $\xi^k = [\xi_1^k, \xi_2^k, \ldots, \xi_n^k] \in \mathcal{W}$, and the best solution all the particles have found $\zeta = [\zeta_1, \zeta_2, \ldots, \zeta_n] \in \mathcal{W}$. Notice that the superscript $k$ is not a power index, but a symbol referring a variable for the $k$th particle $b_k$. The behavior of the particles is controlled by the updating equations of the instantaneous velocity and position. In this study, during each iteration, particle velocity is first updated, followed by the updating of particle position. For the $k$th particle $b_k, k = 1, 2, \ldots, m$, the $i$th element of the velocity $v^k$ is updated in a two-step procedure

$$\text{Step 1: } \gamma = 0.729 \times \{v_i^k(t) + 2.8\rho_1[\xi_i^k(t) - \alpha_i^k(t)] + 1.3\rho_2[\zeta_i(t) - \alpha_i^k(t)]\}$$

$$\text{Step 2: } v_i^k(t+1) = \begin{cases} \alpha_i^{\min} - \alpha_i^{\max} & \text{, if } \gamma < \alpha_i^{\min} - \alpha_i^{\max} \\ \alpha_i^{\max} - \alpha_i^{\min} & \text{, if } \gamma > \alpha_i^{\max} - \alpha_i^{\min} \\ \gamma & \text{, otherwise,} \end{cases} \quad (4.1)$$

where $i = 1, 2, \ldots, n$, $\rho_1$ and $\rho_2$ are two random variables within $[0, 1]$, and $t$ is the iteration number. All the constants in Step 1 are obtained from [14]. Step 2 limits the velocity within a certain range. With the new velocity, the particle position is obtained through another two-step procedure:

$$\text{Step 1: } \gamma = \alpha_i^k(t) + v_i^k(t + 1)$$

$$\text{Step 2: } \alpha_i^k(t + 1) = \begin{cases} \alpha_i^{\min} & \text{, if } \gamma < \alpha_i^{\min} \\ \alpha_i^{\max} & \text{, if } \gamma > \alpha_i^{\max} \\ \gamma & \text{, otherwise ,} \end{cases}$$

(4.2)

where $i = 1, 2, \ldots, n$ and $k = 1, 2, \ldots, m$. Step 2 guarantees that the particles are located within the parametric space $\mathcal{W}$. If a particle hits the boundary of $\mathcal{W}$ in a specific dimension after its position is updated, it will be reflected back by changing the direction of the velocity in that dimension

$$v_i^k = -v_i^k, \text{ if } \alpha_i^k = \alpha_i^{\min} \text{ or } \alpha_i^{\max}$$

(4.3)

where $i = 1, 2, \ldots, n$ and $k = 1, 2, \ldots, m$. Thus, the particle will not be trapped on the boundary of $\mathcal{W}$.

Equations (4.1)–(4.3) provide a skeleton of the PSO method. For a specific application, definitions of $\alpha^{\min}$, $\alpha^{\max}$, and the fitness function $\mathcal{F}$ are required.

### 4.2.2 PSO MODEL FOR GETTING A NARROWBAND DIPOLE MODEL

An electric or magnetic Hertzian dipole is defined by seven parameters, dipole position ($x'$, $y'$, $z'$), dipole direction ($\phi'$, $\theta'$), and the magnitude ($\chi'$) as well as phase ($\psi'$) of the dipole moment. All these dipole parameters comprises a parametric space $\mathcal{W}$ with $7N$ dimensions if $N$ dipoles are adopted. $\alpha^{\min}$ and $\alpha^{\max}$ that are required by the PSO method are determined by the ranges of the dipole parameter as given in Table 4.1. $\psi' \in (\pi, 2\pi)$ is excluded because a dipole with parameters ($x'$, $y'$, $z'$, $\phi'$, $\theta'$, $\chi'$, $\psi'$), where $\psi' \in (\pi, 2\pi)$, can also be represented by parameters ($x'$, $y'$, $z'$, $\phi' + \pi$, $\pi - \theta'$, $\chi'$, $\psi' - \pi$), where ($\psi' - \pi) \in [0, \pi]$. $\chi^{\max}$ is chosen to be 1 A·m for an electric Hertzian dipole, and $\eta_0$ V·m for a magnetic Hertzian dipole, in order to make the radiation from these two types of dipoles comparable. $\eta_0$ is the characteristic impedance of free space.

Table 4.1: Ranges of dipole parameters; from [23], copyright © 2008 IEEE.

| | $x'$ | $y'$ | $z'$ | $\phi'$ | $\theta'$ | $\chi'$ | $\psi'$ |
|---|---|---|---|---|---|---|---|
| Minimum | $x^{\min}$ | $y^{\min}$ | $z^{\min}$ | 0 | 0 | 0 | 0 |
| Maximum | $x^{\max}$ | $y^{\max}$ | $z^{\max}$ | $2\pi$ | $\pi$ | $\chi^{\max}$ | $\pi$ |

The fitness function $\mathcal{F}$ that is also required by an optimization algorithm is the root mean square (RMS) error of the near-field data defined as

$$\varepsilon_{\text{nb}} = \sqrt{\frac{1}{2s} \sum_{(x,y,z)_1}^{(x,y,z)_s} \left| \frac{\vec{E}^e(x, y, z)}{P^e} - \frac{\vec{E}^d(x, y, z)}{P^d} \right|^2},$$

$$\text{where } P^{e,d} = \sum_{(x,y,z)_1}^{(x,y,z)_s} \left| \vec{E}^{e,d}(x, y, z) \right|^2. \tag{4.4}$$

In (4.4), $\vec{E}^e(x, y, z)$, obtained by using WIPL-D, is the exact tangential electric field on an observation surface enclosing the DRA, and $\vec{E}^d(x, y, z)$ is that at the same location but due to the dipoles and their images. $s$ is the index of the sample position. The term $\sqrt{1/2}$ exists in (4.4) because the RMS error is computed for the two orthogonal components of all the tangential electric field data. Note that in (4.4), the near-field data are normalized before computing the error. This is because the dipole model is used to recover the spatial distribution of the near field but not the exact value that also depends on excitation. On the other hand, the exact field due to the excitation of $V^+$ can be obtained by multiplying all the dipole moments with a factor of $(V^+ P^e)/(V_0^+ P^d)$ where $V_0^+$ is the antenna excitation used to get the exact near-field data in (4.4). Furthermore, with this normalization, the optimum solution in the sense of minimizing (4.4) is a volume but not a point in the parametric space which may result in fast convergence in optimization.

As examples, the three dipole models listed in Tables 2.2–2.4 are obtained using the procedure described above.

## 4.3    LIMITATIONS OF THE NARROWBAND MODEL

The narrowband model described above is obtained by an optimization algorithm such that the near-field data at an operating frequency due to the dipoles are very close to those due the physical antenna. It is valid only within a narrow frequency band as will be shown later. In [22], a wideband dipole model is presented by multiplying a frequency dependent scaling factor with the dipole moments while fixing all other dipole parameters so that the near-field variation with frequency due to the antenna matching is taken into account. It was pointed out in [22] that the wideband model is valid within a frequency band where no new mode is excited. However, the requirement of fixed-mode operation is normally satisfied only for a narrowband antenna, and is often violated for a wideband antenna where multiple modes are excited. Therefore, the wideband model given in [22] is able to capture the radiation characteristics of a narrowband antenna within a wide band, including both the bands with and without good matching, but may fail to model a wideband antenna within its operating band. This limitation can be demonstrated by studying the example below.

Table 4.2: Hertzian Dipole Model 1 for the stacked DRA at 9.0 GHz, with 2.66% RMS error.

| Type | $x\hat{a}L^2_{J,M}$, m | $y\hat{a}L^2_{J,M}$, m | $z'_{J,M}$, m | $\theta'_{J,M}$, rad | $\phi'_{J,M}$, rad | Moment |
|---|---|---|---|---|---|---|
| J | 7.05831e-4 | 2.01059e-3 | 2.15846e-3 | 1.21785 | 5.42256 | 0.689837exp(j1.13855) AÂ·m |
| J | 2.66094e-3 | 2.78484e-4 | 2.70365e-3 | 3.04207 | 2.85986 | 1.09291exp(j2.06104) AÂ·m |
| J | -2.47082e-3 | -1.8634e-3 | 6.7026e-4 | 1.83539 | 2.90116 | 1.25955exp(j1.39082) AÂ·m |
| J | -1.60546e-3 | -4.72949e-4 | 3.2831e-3 | 2.3612 | 6.15796 | 0.628064exp(j0.25607) AÂ·m |
| M | -2.0435e-3 | -2.19817e-3 | 4.385e-3 | 1.53652 | 1.38868 | 38.9827exp(j0.674369) VÂ·m |
| M | -1.04419e-3 | 1.70643e-3 | 3.44554e-3 | 1.70399 | 1.62693 | 137.606exp(j1.72249) VÂ·m |
| M | -5.28206e-4 | -7.13482e-4 | 2.29657e-3 | 1.3964 | 1.75371 | 366.081exp(j2.44707) VÂ·m |

Table 4.3: Hertzian Dipole Model 2 for the stacked DRA at 9.0 GHz, with 2.70% RMS error.

| Type | $x\hat{a}L^2_{J,M}$, m | $y\hat{a}L^2_{J,M}$, m | $z'_{J,M}$, m | $\theta'_{J,M}$, rad | $\phi'_{J,M}$, rad | Moment |
|---|---|---|---|---|---|---|
| J | -1.60716e-3 | 2.14792e-4 | 4.437e-3 | 1.67416 | 0.0564627 | 0.529923exp(j0.282972) AÂ·m |
| J | 2.48681e-3 | 2.30593e-4 | 2.83305e-4 | 1.99007 | 5.82041 | 0.897734exp(j1.27304) AÂ·m |
| J | 1.84223e-3 | -4.39336e-3 | 2.90159e-3 | 3.08949 | 5.39214 | 0.264892exp(j1.27279) AÂ·m |
| J | 4.42893e-3 | 2.19881e-4 | 2.01977e-3 | 2.87729 | 0.718597 | 0.835271exp(j2.0199) AÂ·m |
| M | -3.69866e-4 | 7.3468e-4 | 2.1887e-3 | 1.26346 | 1.85508 | 357.870exp(j1.76488) VÂ·m |
| M | 4.6933e-3 | -9.74001e-4 | 1.58392e-3 | 0.461097 | 5.53025 | 47.3481exp(j2.5372) VÂ·m |
| M | 1.2006e-3 | 1.99774e-3 | 2.93881e-3 | 1.65447 | 6.07813 | 85.7907exp(j0.907767) VÂ·m |

A probe-fed DRA above an infinite ground plane is illustrated in Figure 4.1, where the probe is tapered, with its radius linearly decreasing from 0.3 mm at the bottom to 0.2 mm at the top. It is simulated by the Method of Moments software package WIPL-D [13], and its −10 dB bandwidth is approximately from 7.67 to 11 GHz. Two different dipole models, namely Model 1 as given in Table 4.2 and Model 2 as given in Table 4.3 where each model has four electric and three magnetic Hertzian dipoles, are obtained to model the DRA at 9 GHz by minimizing (4.4).

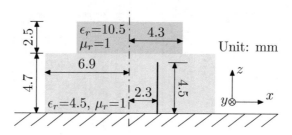

Figure 4.1:  Stacked DRA; from [23], copyright © 2008 IEEE.

In this example, the observation surface consists of five surfaces of a box defined by $x = \pm 15$ mm, $y = \pm 15$ mm, and $z = 20$ mm. A total of 100 observation points are evenly sampled at each surface. The RMS errors of the dipole models are 2.66% for Model 1 and 2.70% for Model 2. Based on Uniqueness Theorem, either dipole model can be used to represent the physical DRA in the sense of generating similar field. The radiation patterns of the two dipole models are plotted against those of the physical DRA in Figure 4.2, where perfect agreement is observed. If either Model 1 or Model 2 is used to represent the DRA from 7.5 to 11 GHz, the resulting RMS errors are plotted in Figure 4.3 where the errors are small only around 9 GHz. In [22], a scaling factor

$$\gamma(f) = \frac{Z_{\text{in}}(f_0)}{Z_{\text{in}}(f)} , \tag{4.5}$$

where $Z_{\text{in}}(f)$ is the antenna input impedance at a frequency $f$ and $f_0$ is the frequency at which the dipole model is obtained, is introduced to the moment of each dipole while fixing all the other dipole parameters in order to get a wideband model. However, this procedure may not give a satisfactory solution for some types of antennas. For example, if Model 1 or Model 2 is used to construct a wideband model for the DRA in Figure 4.1 from 7.5 to 11 GHz following the above procedure, the resulting RMS errors are plotted in Figure 4.4, where the error quickly increases when the operating frequency goes away from 9.0 GHz. Clearly, the dipole model obtained from the near-field data at a single frequency is only valid over a very narrow band for this wideband antenna.

An idea to model the antenna over a wide bandwidth is to use the interpolation technique assuming that a small disturbance of the operating frequency results in a small disturbance of the dipole parameters. With several single-frequency dipole models separately obtained at different frequencies within the band of interest, it is expected that the dipole model at any frequency within

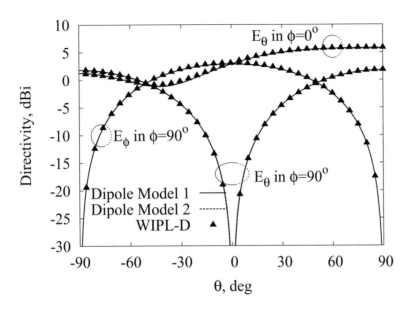

**Figure 4.2:** Radiation patterns of the stacked DRA at 9 GHz; from [23], copyright © 2008 IEEE.

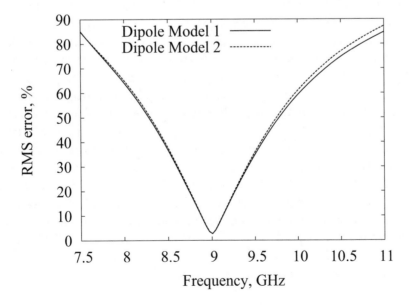

**Figure 4.3:** RMS error $\varepsilon_{nb}$ of the narrowband model

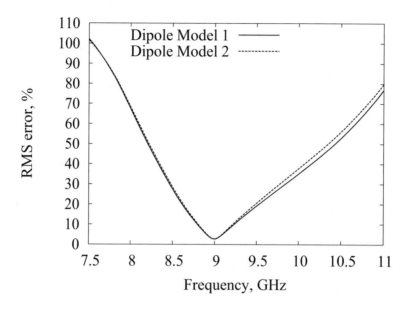

**Figure 4.4:** RMS error $\varepsilon_{nb}$ of the wideband model given in [22]; from [23], copyright © 2008 IEEE.

that band can be interpolated. Furthermore, an analytical expression in terms of frequency for each dipole parameter may also be extracted by using curve fitting. Unfortunately, besides the inconvenient multiple single-frequency models, the idea of interpolation does not work because the dipole model solution is not unique. Suppose there are two different dipole models for a given antenna at a frequency $f$, one with the dipole parameter vector $\mathbf{A}$ and the other with the parameter vector $\mathbf{B}$. Then, at a frequency $f + \Delta f$, the dipole parameters should be disturbed to be $\mathbf{A} + \Delta \mathbf{A}$ or $\mathbf{B} + \Delta \mathbf{B}$. The interpolation of $\mathbf{A}$ and $\mathbf{A} + \Delta \mathbf{A}$, or $\mathbf{B}$ and $\mathbf{B} + \Delta \mathbf{B}$ will give a valid dipole model between the frequencies $f$ and $f + \Delta f$. However, there is no way to guarantee that the separately obtained dipole models at two frequencies are the disturbed versions of each other. Specifically, if $\mathbf{A}$ is obtained at the frequency $f$ and $\mathbf{B} + \Delta \mathbf{B}$ is obtained at the frequency $f + \Delta f$, the interpolation will fail. This can be shown with the two obtained dipole models for the DRA in Figure 4.1. As a special case of interpolation, the average of the dipole parameters in Model 1 and Model 2 should give another valid dipole model at 9 GHz if the idea of interpolation works. However, the averaged model gives an RMS error of 99.9% at 9 GHz, and its radiation patterns as shown in Figure 4.5 are completely different to those in Figure 4.2. Therefore, it is desirable to obtain a wideband Hertzian dipole model simultaneously using the near-field data at different frequencies.

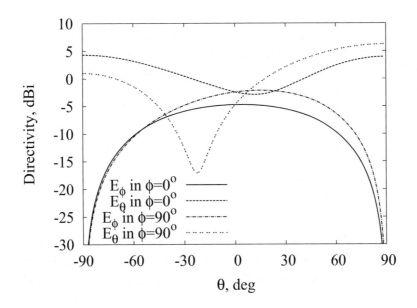

**Figure 4.5:** Radiation patterns of the interpolated dipole model; from [23], copyright © 2008 IEEE.

## 4.4    WIDEBAND HERTZIAN DIPOLE MODEL

### 4.4.1    PSO MODEL FOR GETTING A WIDEBAND DIPOLE MODEL

In [23], a wideband dipole model is proposed by making each dipole parameter a $(p-1)^{\text{th}}$ order polynomial in the free space phase constant $\beta$ as:

$$
\begin{aligned}
x'(\beta) &= C_{p-1}^{x}\beta^{p-1} + C_{p-2}^{x}\beta^{p-2} + \cdots + C_{1}^{x}\beta + C_{0}^{x} \\
y'(\beta) &= C_{p-1}^{y}\beta^{p-1} + C_{p-2}^{y}\beta^{p-2} + \cdots + C_{1}^{y}\beta + C_{0}^{y} \\
z'(\beta) &= C_{p-1}^{z}\beta^{p-1} + C_{p-2}^{z}\beta^{p-2} + \cdots + C_{1}^{z}\beta + C_{0}^{z} \\
\phi'(\beta) &= C_{p-1}^{\phi}\beta^{p-1} + C_{p-2}^{\phi}\beta^{p-2} + \cdots + C_{1}^{\phi}\beta + C_{0}^{\phi} \;, \\
\theta'(\beta) &= C_{p-1}^{\theta}\beta^{p-1} + C_{p-2}^{\theta}\beta^{p-2} + \cdots + C_{1}^{\theta}\beta + C_{0}^{\theta} \\
\chi'(\beta) &= C_{p-1}^{\chi}\beta^{p-1} + C_{p-2}^{\chi}\beta^{p-2} + \cdots + C_{1}^{\chi}\beta + C_{0}^{\chi} \\
\psi'(\beta) &= C_{p-1}^{\psi}\beta^{p-1} + C_{p-2}^{\psi}\beta^{p-2} + \cdots + C_{1}^{\psi}\beta + C_{0}^{\psi}
\end{aligned}
\tag{4.6}
$$

where $p = 1, 2, \ldots$. The $\beta$ of interest is within $[\beta^{\min}, \beta^{\max}]$, where $\beta^{\min}$ and $\beta^{\max}$ are the free space phase constants for the lowest and highest frequencies, respectively. The value of each dipole parameter has a range within the band of $[\beta^{\min}, \beta^{\max}]$, as listed in Table 4.1.

The optimization objective is to determine all the polynomial coefficients in (4.6). However, without knowledge of the ranges of these coefficients, they cannot be selected to construct the optimization parametric space $\mathcal{W}$ because $\alpha^{\min}$ and $\alpha^{\max}$, which are required to define $\mathcal{W}$ and for the PSO updating procedure (4.1)–(4.3), are unknown. Instead, for each dipole parameter, the values of its polynomial at $p$ different frequencies are selected to construct $p$ dimensions of $\mathcal{W}$. Take $x'$ for example; the optimization parameters are selected to be $x'(\beta_i)$, where $\beta_i = \beta^{\min} + i(\beta^{\max} - \beta^{\min})/(p-1)$ for $i = 0, 1, 2, \ldots, p-1$. These variables can be selected to construct the parametric space $\mathcal{W}$ because their ranges are predefined as $[x^{\min}, x^{\max}]$ in Table 4.1. After updating $x'(\beta_i)$ for $i = 0, 1, \ldots, p-1$ during each PSO iteration, the new polynomial coefficients for $x'(\beta)$ can be obtained as

$$
\begin{bmatrix} C^x_{p-1} \\ C^x_{p-2} \\ \vdots \\ C^x_0 \end{bmatrix} = \begin{bmatrix} \beta_0^{p-1} & \beta_0^{p-2} & \cdots & 1 \\ \beta_1^{p-1} & \beta_1^{p-2} & \cdots & 1 \\ \vdots & \vdots & \vdots & \vdots \\ \beta_{p-1}^{p-1} & \beta_{p-1}^{p-2} & \cdots & 1 \end{bmatrix}^{-1} \begin{bmatrix} x'(\beta_0) \\ x'(\beta_1) \\ \vdots \\ x'(\beta_{p-1}) \end{bmatrix}. \tag{4.7}
$$

Similarly, $y'(\beta_i)$, $z'(\beta_i)$, $\phi'(\beta_i)$, $\theta'(\beta_i)$, $\chi'(\beta_i)$, and $\psi'(\beta_i)$, for $i = 0, 1, \ldots, n-1$, are selected as other dimensions of $\mathcal{W}$. The coefficients of these dipole parameters can be obtained in a similar manner as in (4.7). Given $N_J$ electric and $N_M$ magnetic Hertzian dipoles to model an antenna, $\alpha^{\min}$ and $\alpha^{\max}$, both with a length of $n = 7p(N_J + N_M)$, can be defined with the data in Table 4.1. In either $\alpha^{\min}$ or $\alpha^{\max}$, there are $7p$ elements for each electric or magnetic Hertzian dipole, with 7 elements for each of the $p$ frequencies.

The fitness function $\mathcal{F}$, also required for the implementation of PSO, is defined in (4.8) below as the RMS error of the electric field generated by a set of Hertzian dipoles when compared to exact field, which is normally obtained from a full wave analysis of the real antenna.

$$
\mathcal{F} = \varepsilon_{\text{wb}} = \sqrt{\frac{1}{2s} \sum_{(x,y,z,\beta)_1}^{(x,y,z,\beta)_s} \left| \frac{\vec{E}^e(x, y, z, \beta)}{P^e} - \frac{\vec{E}^d(x, y, z, \beta)}{P^d} \right|^2},
$$

$$
\text{where } P^{e,d} = \sum_{(x,y,z,\beta)_1}^{(x,y,z,\beta)_s} \left| \vec{E}^{e,d}(x, y, z, \beta) \right|^2. \tag{4.8}
$$

In (4.8), $\vec{E}^e(x, y, z, \beta)$ is the exact electric field at an observation point of $(x, y, z)$ and at a frequency with the free space phase constant $\beta$, where the observation point is located on a surface enclosing the real antenna, and $\vec{E}^d(x, y, z, \beta)$ has a similar notation, but for the field generated by the dipoles. A total of $s$ near-field samples at different positions or frequencies are used. A set of frequency dependent dipoles can be synthesized by minimizing $\varepsilon_{\text{wb}}$. If the dipoles are in free space, $\vec{E}^d(x, y, z, \beta)$ is derived in Chapter 6 of [24]. If the dipoles are above an infinite ground plane, the radiation due the dipole images should also be included.

With $\alpha^{\min}$ as well as $\alpha^{\max}$ defined in Table 4.1, and the fitness function defined in (4.8), the PSO method in (4.1)–(4.3) can be applied to get the Hertzian dipole model for an antenna.

## 4.4.2    MODELING OF A WIDEBAND ANTENNA

As an example, the wideband DRA in Figure 4.1 is modeled by a set of frequency dependent Hertzian dipoles from 7.5–11 GHz. First, the tangential electric field on a surface enclosing the DRA is obtained by using WIPL-D. The surface and the sample positions are chosen to be the same as those described in Section 4.3. At each sample position, field data at 8 frequencies, from 7.5–11 GHz, with a 0.5 GHz increment, are obtained to evaluate the fitness function in (4.8). Then, the method presented above is used to determine the coefficients of the dipole parameters. $x^{\min}$, $x^{\max}$, $y^{\min}$, $y^{\max}$, $z^{\min}$, and $z^{\max}$ required to define $\alpha^{\min}$, and $\alpha^{\max}$ are given in Table 4.4.

Table 4.4:  Ranges of dipole positions; from [23], copyright © 2008 IEEE.

| $x^{\min}$ | $x^{\max}$ | $y^{\min}$ | $y^{\max}$ | $z^{\min}$ | $z^{\max}$ |
|---|---|---|---|---|---|
| −8 mm | 8 mm | −8 mm | 8 mm | 0 mm | 8 mm |

Before applying the optimization algorithm, the numbers of the electric and magnetic dipoles should be chosen first. The dipole number, which depends on the spatial variation of the near-field data, can be determined by examining the required number of dipoles at the upper frequency of 11 GHz because the field varies the most at the upper frequency. Figure 4.6 shows the convergence curves for six narrowband dipole models at 11 GHz. It takes about 6 min on a processor of 1.87 GHz to get each curve. It can be seen that the model with four electric and three magnetic dipoles and that with five electric and two magnetic dipoles give the smallest error at 3%. Therefore, either combination of the dipoles is required for a wideband model from 7.5–11 GHz. More dipoles may give better results, but it also increases the parametric space and requires more time for each optimization iteration. Herein, four electric and three magnetic Hertzian dipoles are chosen to model the wideband DRA.

Thirty particles are used to search for the optimum solution $\alpha_{\mathrm{opt}}$ in the parametric space $\mathcal{W}$. All the dipoles radiate in the presence of the infinite ground plane. A total of four Hertzian dipole models are obtained, with the $p$ in (4.6) from 2 to 5. After 12,000 iterations, the resulting $\varepsilon_{\mathrm{wb}}$ values in (4.8) and the required time on a 64-bit processor of 2.2 GHz are listed in Table 4.5 for different $p$ values, where the models with $p = 3$ or 4 give better results. Although the optimization procedure is

Table 4.5:  RMS errors of the dipole models with different $p$; from [23], copyright © 2008 IEEE.

| $p$ | 2 | 3 | 4 | 5 |
|---|---|---|---|---|
| $\varepsilon_{\mathrm{wb}}$ | 6.93% | 3.94% | 3.89% | 6.52% |

time consuming, once the dipole model is obtained, it can be used to efficiently study the radiation characteristics of the DRA mounted on other platforms as will be discussed in Section 4.4.3. The $\varepsilon_{\mathrm{wb}}$ values in Table 4.5 are calculated with all the near-field data at 8 frequencies from 7.5–11 GHz, and thus provide no information about the error at a single frequency. In order to examine the error

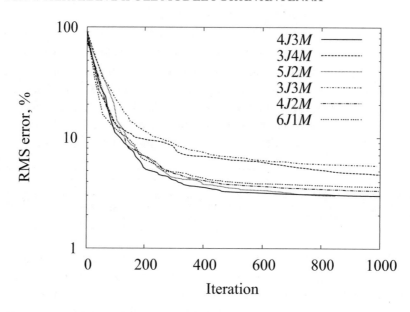

**Figure 4.6:** Convergence plots of six narrowband dipole models at 11 GHz; from [23], copyright © 2008 IEEE.

distribution over the frequency band of interest, (4.4) is evaluated for different frequencies separately, and the results are plotted in Figure 4.7 for all the four dipole models with different polynomial orders. The model with first order polynomials where $p = 2$ assumes linear relationship between the dipole parameters and the frequency, which cannot track the variation of the near field with frequency, and resulted in relatively large error. On the other hand, the models with higher order polynomials such as $p = 4$ or 5 provide too much variation freedom which leads to large error at the frequencies where no near-field data are available for optimization. Therefore, the Hertzian dipole model with second order polynomials where $p = 3$ is the best one for the DRA in Figure 4.1. Normally, the higher the order of polynomials chosen, the more frequencies with near-field data that are desired in the optimization to obtain a good Hertzian dipole model.

After obtaining all the polynomial coefficients as given in Table 4.6 for $p = 3$, the dipole parameters versus frequency are shown in Figure 4.8. It can be seen that some parameters such as $\theta'$ of $M_2$ and $\phi'$ of $J_3$ slightly exceed the predefined ranges at some frequencies, but this has no significant effect on the application of the Hertzian dipole model. Especially, if the $z'$ coordinate of a dipole is negative, which by accident does not happen in Figure 4.8, that dipole is actually a dipole image below the ground plane, and the physical dipole can be obtained by using the image of that dipole. With the wideband dipole model, the far-field radiation pattern can be explicitly computed for any frequency within the band of interest. The radiation patterns at 7.5, 8.25, 9, and 10.75 GHz due

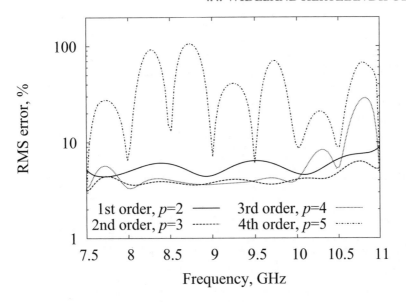

**Figure 4.7:** RMS error $\varepsilon_{nb}$ of the dipole models with different polynomial orders; from [23], copyright © 2008 IEEE.

to the optimal dipole model are compared to the exact ones obtained from WIPL-D in Figure 4.9, where perfect agreement is observed. Although not shown here, good agreement is also observed at other frequencies, both frequencies at which the near-field data were used in the optimization and those at which the near-field data were not used. Therefore, the set of frequency dependent Hertzian dipoles is a good model for the wideband stacked DRA, in the sense of generating similar radiation in both the near and far field regions.

The location of the enclosing box where the near-field data are sampled and the number of samples may affect the quality of a dipole model and the convergence of the optimization procedure. For the stacked DRA, if the sampling box shrinks to be $x = \pm 10$ mm, $y = \pm 10$ mm, and $z = 10$ mm, it requires 1,800 iterations to achieve a 3% RMS error for a model with 4 electric and 3 magnetic dipoles at 11 GHz while it only needs 1,000 iterations to get the same error for the larger box used before. The sampling number should also be large enough to capture the spatial variation of the near field. This can be determined using Nyquist's sampling theorem after studying the spatial spectrum of the near field on the sampling box. Specifically, the more directive an antenna is, the greater the near field varies and thus more samples are required.

### 4.4.3 APPLICATION

The dipole model in Section 4.4.2 represents the DRA over an infinite ground plane. This model can be used to efficiently investigate the radiation characteristics of the DRA mounted on other

Table 4.6: Polynomial coefficients for the optimal Hertzian dipole parameters, $p = 3$.

| | $J_1$ | $J_2$ | $J_3$ | $J_4$ | $M_1$ | $M_2$ | $M_3$ |
|---|---|---|---|---|---|---|---|
| $C_2^x$ | -3.94333e-7 | -1.39720e-6 | 1.33250e-6 | -9.78851e-7 | -9.61397e-7 | -1.17535e-6 | 2.56412e-6 |
| $C_1^x$ | 1.56201e-4 | 5.28508e-4 | -3.98580e-4 | 2.94732e-4 | 3.33758e-4 | 3.88452e-4 | -1.14636e-3 |
| $C_0^x$ | -1.55391e-2 | -4.59769e-2 | 2.79312e-2 | -1.83430e-2 | -2.92617e-2 | -2.95884e-2 | 1.24829e-1 |
| $C_2^y$ | 1.43975e-6 | -4.58313e-9 | 2.24458e-7 | 9.88262e-7 | -5.36477e-7 | 1.52089e-6 | 2.57408e-7 |
| $C_1^y$ | -5.37955e-4 | -2.63465e-5 | -6.65552e-5 | -3.63180e-4 | 1.99733e-4 | -5.97392e-4 | -1.45694e-4 |
| $C_0^y$ | 4.88006e-2 | 5.25170e-3 | 4.71890e-3 | 3.27347e-2 | -1.85960e-2 | 5.77805e-2 | 1.77017e-2 |
| $C_2^z$ | 3.96693e-7 | 1.00131e-6 | -1.13407e-6 | 1.83660e-6 | 1.56104e-8 | 1.15043e-7 | 1.67087e-6 |
| $C_1^z$ | -1.32922e-4 | -4.67075e-4 | 4.32270e-4 | -6.98869e-4 | 1.37207e-5 | -5.33700e-5 | -7.28210e-4 |
| $C_0^z$ | 1.23973e-2 | 5.62707e-2 | -3.59515e-2 | 6.85332e-2 | -1.60122e-3 | 6.71577e-3 | 8.10940e-2 |
| $C_2^\theta$ | -1.44574e-4 | -3.39951e-4 | 4.70239e-4 | -1.37846e-4 | -1.23908e-4 | 1.21526e-3 | 7.11444e-4 |
| $C_1^\theta$ | 4.08721e-2 | 1.53694e-1 | -1.49924e-1 | 2.69213e-2 | 4.12948e-2 | -4.34360e-1 | -2.86426e-1 |
| $C_0^\theta$ | 1.05327e-2 | -1.42232e+1 | 1.25951e+1 | 2.30854e+0 | -1.53697e+0 | 3.86893e+1 | 2.97675e+1 |
| $C_2^\phi$ | -1.14388e-3 | -1.13113e-3 | 5.65382e-5 | 2.02684e-5 | -1.78792e-4 | -1.06640e-3 | -1.16937e-3 |
| $C_1^\phi$ | 4.90185e-1 | 3.84570e-1 | -2.31240e-2 | 2.73565e-3 | 6.53568e-2 | 4.44773e-1 | 4.94002e-1 |
| $C_0^\phi$ | -4.65530e+1 | -2.67041e+1 | 2.35858e+0 | 1.72503e+0 | -4.29061e+0 | -4.17552e+1 | -4.76617e+1 |
| $C_2^\chi$ | -3.32402e-5 | -1.36534e-4 | 1.39920e-4 | 8.08284e-6 | -1.57825e-2 | -1.96023e-2 | 1.24683e-2 |
| $C_1^\chi$ | 2.26679e-3 | 5.65019e-2 | -6.29484e-2 | -1.23585e-2 | 4.01300e+0 | 5.70288e+0 | -2.70504e+0 |
| $C_0^\chi$ | 1.45280e+0 | -5.27986e+0 | 7.19957e+0 | 2.70846e+0 | 1.35729e+2 | -2.17435e+2 | 1.28630e+2 |
| $C_2^\psi$ | 4.31504e-4 | -1.16043e-5 | 1.09266e-3 | -1.34906e-4 | 3.24276e-4 | -8.13079e-5 | -2.02993e-4 |
| $C_1^\psi$ | -1.83993e-1 | -9.41356e-3 | -4.48638e-1 | 3.86885e-2 | -1.44482e-1 | 3.50834e-3 | 6.58134e-2 |
| $C_0^\psi$ | 1.99238e+1 | 3.88077e+0 | 4.66645e+1 | -7.68714e-3 | 1.78159e+1 | 4.58664e+0 | -2.19657e+0 |

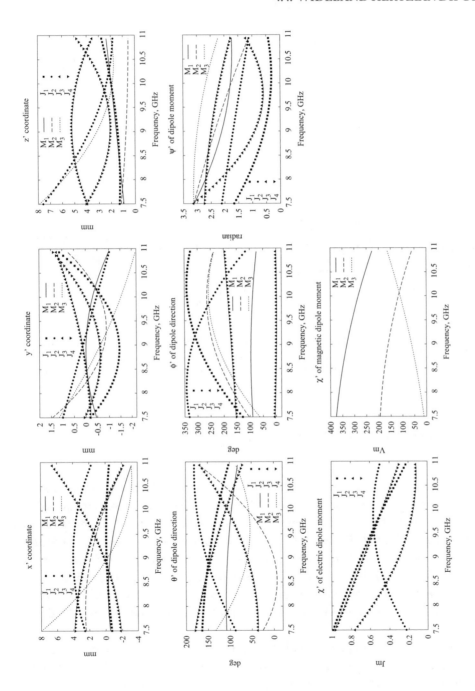

**Figure 4.8:** Variation of the optimal dipole parameters with frequency, $p = 3$; from [23], copyright © 2008 IEEE.

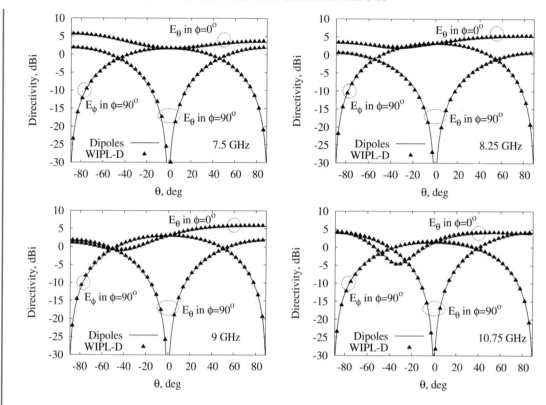

**Figure 4.9:** Radiation patterns of the DRA; from [23], copyright © 2008 IEEE.

platforms assuming that the difference between the platforms has little effect on the dipole model. In Chapters 2 and 3, three narrowband dipole models listed in Tables 2.2–2.4 were applied to multilayer dielectric structures. Herein, a different platform is chosen for the wideband model of the stacked DRA. As shown in Figure 4.10, the DRA is mounted at three different positions of a circular ground plane. The $S_{11}$ of the DRA at different locations are plotted against that for the infinite ground plane case in Figure 4.11. It can be seen that the antenna matching is not sensitive to the platform because it is mostly determined by the neighboring environment of the DRA. However, the radiation pattern may be greatly affected by the platform and can be studied using the dipole model as follows.

The DRA in Figure 4.10 is first replaced by the dipole model, and then simulated by using a Method of Moments software package FEKO [25] that is capable of incorporating Hertzian dipoles. The radiation patterns obtained from FEKO are compared to those obtained from a full wave analysis of the physical structure using HFSS in Figure 4.12 for 8.25 GHz and in Figure 4.13 for 11 GHz. Good agreement is observed for all the cases. Note that although there is no near-field data at 8.25

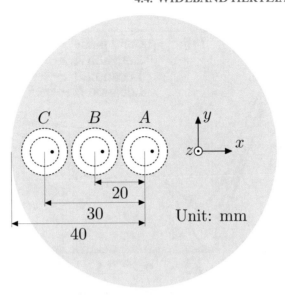

**Figure 4.10:** A DRA mounted at different positions of a circular ground plane; from [23], copyright ©
2008 IEEE.

GHz used in the optimization, the dipole model still captures the radiation characteristics of the
DRA on the finite ground plane at that frequency. Although not shown here, the agreement at other
frequencies is also observed. It is understandable that the dipole model will be valid for a larger
ground plane since it is obtained for the DRA on an infinite ground plane. It should be emphasized
that the dipole model is valid only for a platform that does not significantly disturb the antenna's
operating mode. Otherwise, a full wave analysis of the entire structure should be applied.

The advantages of using a dipole model to study the structure in Figure 4.10 over a full wave
analysis of the entire structure can be demonstrated in a Method of Moments framework. In the
Method of Moments, a linear system is constructed and solved at each frequency as

$$ZI = V \implies I = Z^{-1}V \,, \tag{4.9}$$

where $Z$ is the impedance matrix determined by the structure geometry, $I$ is the unknown vector,
and $V$ is the excitation vector. For a full wave analysis of the physical structure, the change of the
DRA location changes part of the matrix $Z$ and thus $Z^{-1}$ is required to be recomputed. However,
if the DRA is replaced by the dipole model, the matrices $Z$ and $Z^{-1}$ are fixed for different DRA
locations and the recomputing of $Z^{-1}$ is avoided. In addition, the system order reduces because the
unknowns associated with the DRA are replaced by the known dipole model. This concept can be
easily extended to study an antenna mounted on other types of complex platforms, such as the body
of a car, ship, or airplane. In these applications, the $Z$ matrix is very large, and the avoidance of
recomputing $Z^{-1}$ for different antenna locations saves a lot of computation time.

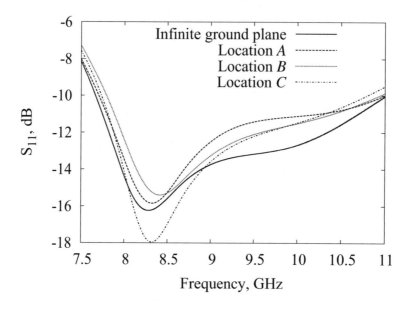

**Figure 4.11:** A DRA mounted at different positions of a circular ground plane; from [23], copyright ©
2008 IEEE.

### 4.4.4    REJECTION OF GAUSSIAN NOISE

The near-field data used to get the Hertzian dipole model may instead be obtained from mea-
surement, and thus may be corrupted with noise. The effect of such noise on the accuracy of the
Hertzian dipole model is investigated. The noises added to the real and imaginary parts of sampled
near-field data are assumed to be independently identical distributed (iid) random variables having
a normal distribution $\mathcal{N}(0, \sigma^2)$. This kind of noise is referred as Gaussian noise. A random variable
$X \sim \mathcal{N}(0, \sigma^2)$ has the probability density function (pdf) of

$$f_X(x) = \frac{1}{\sigma\sqrt{2\pi}} \exp\left(-\frac{x^2}{2\sigma^2}\right), x \in (-\infty, +\infty) . \tag{4.10}$$

From probability theory, the variance $E\{[X - E(X)]^2\} = \sigma^2$, where $E(\cdot)$ calculates the expectation
of a random variable. Therefore, $\sigma$ can be considered as the effective value of the Gaussian noise.
The relative noise level $\kappa$ is defined as the ratio of $\sigma$ and the RMS of the exact near-field data

$$\kappa = \frac{\sigma}{\sqrt{\dfrac{1}{4s} \displaystyle\sum_{(x,y,z,\beta)_1}^{(x,y,z,\beta)_s} \left|\vec{E}^e(x, y, z, \beta)\right|^2}} , \tag{4.11}$$

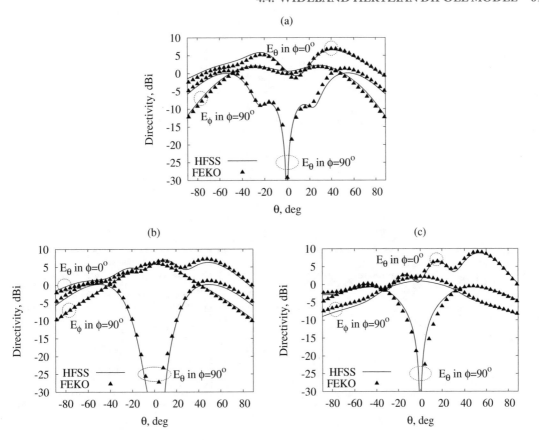

**Figure 4.12:** Radiation patterns of the structure in Figure 4.10 at 8.25 GHz. (a) DRA at position $A$, (b) DRA at position $B$, (c) DRA at position $C$; from [23], copyright © 2008 IEEE.

where the term $\sqrt{1/4}$ exists in the denominator because the Gaussian noise is added to both the real and imaginary parts of the near-field data, for both the two orthogonal components of the tangential electric field.

Different relative noise levels are selected to corrupt the exact near-field data obtained from WIPL-D, and the noise corrupted data are then used to get the Hertzian dipoles. All the parameters used in the PSO method are the same as those in the previous case with $p = 3$ and without noise. After 12,000 iterations, the RMS errors in (4.8) are listed as $\varepsilon_{wb}^{noise}$ in the first row of Table 4.7 for different $\kappa$ values. Furthermore, for each set of Hertzian dipoles associated with different $\kappa$ values, the RMS error in (4.8) is re-evaluated with the exact near-field data that are not corrupted with the Gaussian noise, and listed as $\varepsilon_{wb}^{exact}$ in the second row of Table 4.7. It can be seen that with noise level increasing, $\varepsilon_{wb}^{noise}$ increases while $\varepsilon_{wb}^{exact}$ remains almost the same, around 4%. In other words,

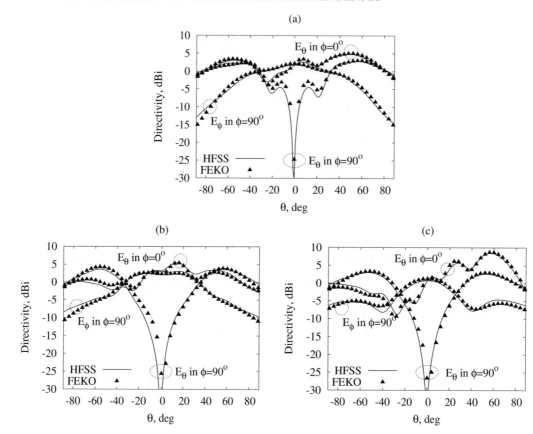

**Figure 4.13:** Radiation patterns of the structure in Figure 4.10 at 11 GHz. (a) DRA at position $A$, (b) DRA at position $B$, (c) DRA at position $C$; from [23], copyright © 2008 IEEE.

the Hertzian dipole model is able to reject the Gaussian noise added to the near-field data, and thus extract the exact field. This is because the Gaussian noise is non-Maxwellian, and has little correlation with the Maxwellian field generated by the Hertzian dipoles. Moreover, for each model

**Table 4.7:** RMS errors of the dipole models that are obtained from noisy data; from [23], copyright © 2008 IEEE.

| $\kappa$ | 20% | 15% | 10% | 5% | 0% |
|---|---|---|---|---|---|
| $\varepsilon_{\text{wb}}^{\text{noise}}$ | 20.01% | 15.24% | 10.73% | 6.58% | 3.94% |
| $\varepsilon_{\text{wb}}^{\text{exact}}$ | 4.25% | 3.97% | 4.11% | 4.39% | 3.94% |

obtained from noisy data, the RMS error $\varepsilon_{\text{nb}}$ in (4.4) is computed at different frequencies using the near-field data not corrupted with the Gaussian noise, and plotted in Figure 4.14, where the RMS errors are observed to be less than 6.6% for all the five dipole models within the frequency band from 7.5–11 GHz. The results in Table 4.7 and Figure 4.14 show that the procedure of obtaining

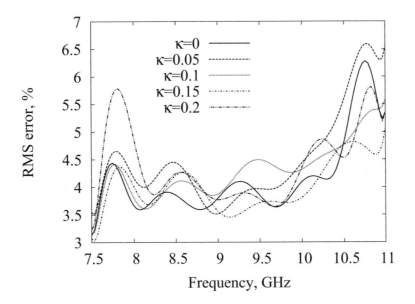

**Figure 4.14:** RMS error $\varepsilon_{nb}$ of the dipole model obtained from Gaussian noise corrupted data, $p = 3$; from [23], copyright © 2008 IEEE.

the Hertzian dipole model is robust in resisting Gaussian noise.

### 4.4.5   FREQUENCY SCALABILITY

Theoretically, an antenna operating within the band of $[\beta^{min}, \beta^{max}]$ can be scaled to operate in the band of $[\beta^{min}/\zeta, \beta^{max}/\zeta]$ by scaling the antenna dimensions with a factor of $\zeta$. This concept can also be used to scale the Hertzian dipole model, and afterwards to model the scaled antenna. The scaling of the dipole model is done by scaling the parameters of each Hertzian dipole. Given a scaling factor $\zeta$, the dipole parameters $x'(\beta)$, $y'(\beta)$, $z'(\beta)$, $\phi'(\beta)$, $\theta'(\beta)$, $\chi'(\beta)$, and $\psi'(\beta)$ are scaled to be $\zeta x'(\zeta\beta)$, $\zeta y'(\zeta\beta)$, $\zeta z'(\zeta\beta)$, $\phi'(\zeta\beta)$, $\theta'(\zeta\beta)$, $\chi'(\zeta\beta)$, and $\psi'(\zeta\beta)$, respectively. As examples, the four Hertzian dipole models obtained in Section 4.4.2 are scaled with $\zeta = 2$, to model a stacked DRA scaled from the one in Figure 4.1 with the same scaling factor. The scaled DRA is simulated by WIPL-D to obtain the exact near-field data. For all the four scaled dipole models, the distributions of the near-field RMS error are examined by using the exact near-field data, and are found to be the same as those shown in Figure 4.7, except that the frequency range is from 3.75–5.5 GHz. Therefore, the Hertzian dipole model can be scaled to model a scaled antenna.

## 4.5  CONCLUSIONS

An optimization procedure was presented to get a set of frequency dependent Hertzian dipoles that model the radiation characteristics of a wideband antenna. All the dipole parameters were polynomials in the phase constant, and the polynomial coefficients were determined by minimizing the difference between the exact or measured near-field data and the field generated by the dipoles at multiple frequencies. The optimization parametric space was constructed by using the values of the dipole parameters at different frequencies. For each point in the parametric space, the polynomial coefficients of a dipole parameter were computed by solving a linear system. The PSO method was applied to search for the optimum solution. As an example, a wideband stacked DRA was modeled by four electric and three magnetic frequency dependent Hertzian dipoles from 7.5–11 GHz. Four dipole models with different polynomial orders were obtained and compared, and the optimal one was validated with WIPL-D, by comparing both the near and far fields. The optimal dipole model was applied to efficiently predict the radiation characteristics of the DRA above a circular ground plane. The model may also be used to replace the DRA mounted on other complex platforms. Furthermore, it was found that the procedure of getting the Hertzian dipole model is able to reject the Gaussian noise, and thus extract the exact electromagnetic field. It was also shown that the Hertzian dipole model is frequency scalable.

# APPENDIX A

# Derivation of Equations in Chapter 2

## A.1 DERIVATION OF EQUATION (2.24)

Figure A.1 illustrates a TLS with a characteristic impedance of $R_c$, length of $l$, and phase constant of $\beta$. The scattering parameters at the input and output ports of the TLS are denoted by $(a_1, b_1)$ and

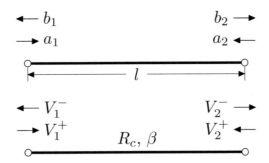

Figure A.1: A transmission line segment.

$(a_2, b_2)$, both obtained for a nominal impedance of 1 $\Omega$, while the forward and backward voltages at the two ports are denoted by $(V_1^+, V_1^-)$ and $(V_2^+, V_2^-)$. The scattering parameters are related to the forward and backward voltages through the relations

$$\begin{cases} a_1 + b_1 = V_1^+ + V_1^- \\ a_1 - b_1 = (V_1^+ - V_1^-)/R_c \\ a_2 + b_2 = V_2^+ + V_2^- \\ a_2 - b_2 = (V_2^+ - V_2^-)/R_c \end{cases} . \tag{A.1}$$

In addition, the backward voltages are dependent on the forward voltages as

$$\begin{cases} V_1^- = V_2^+ \exp(-j\beta\, l) \\ V_2^- = V_1^+ \exp(-j\beta\, l) \end{cases} . \tag{A.2}$$

Thus, $V_1^-$ and $V_2^-$ in (A.1) can be eliminated by substituting (A.2) into (A.1) to obtain

$$
\begin{cases}
a_1 + b_1 = V_1^+ + V_2^+ \exp(-j\beta\, l) \\
a_1 - b_1 = \left[V_1^+ - V_2^+ \exp(-j\beta\, l)\right]/R_c \\
a_2 + b_2 = V_2^+ + V_1^+ \exp(-j\beta l) \\
a_2 - b_2 = \left[V_2^+ - V_1^+ \exp(-j\beta l)\right]/R_c
\end{cases}
\tag{A.3}
$$

$V_1^+$ and $V_2^+$ can be derived from the first two or the last two equations of (A.3), and are given by

$$
\begin{cases}
V_1^+ = (1 + R_c)a_1/2 + (1 - R_c)b_1/2 \\
V_2^+ = [(1 - R_c)a_1/2 + (1 + R_c)b_1/2]\exp(j\beta\, l)
\end{cases}
\tag{A.4}
$$

$$
\begin{cases}
V_1^+ = [(1 - R_c)a_2/2 + (1 + R_c)b_2/2]\exp(j\beta\, l) \\
V_2^+ = (1 + R_c)a_2/2 + (1 - R_c)b_2/2
\end{cases}
\tag{A.5}
$$

Equating the two different representations of both $V_1^+$ and $V_2^+$ in (A.4) and (A.5) gives two linear equations as

$$
\begin{cases}
(1 + R_c)a_1 + (1 - R_c)b_1 = [(1 - R_c)a_2 + (1 + R_c)b_2]\exp(j\beta\, l) \\
[(1 - R_c)a_1 + (1 + R_c)b_1]\exp(j\beta\, l) = (1 + R_c)a_2 + (1 - R_c)b_2
\end{cases}
\tag{A.6}
$$

which may be solved to obtain the final S chain matrix

$$
\begin{bmatrix} a_1 \\ b_1 \end{bmatrix}
= \frac{1}{4R_c} \cdot
\begin{bmatrix}
G R_c + W(1 + R_c^2) & W(1 - R_c^2) \\
W(R_c^2 - 1) & G R_c - W(1 + R_c^2)
\end{bmatrix}
\begin{bmatrix} b_2 \\ a_2 \end{bmatrix},
\tag{A.7}
$$

where $G = 4\cos(\beta l)$ and $W = j2\sin(\beta l)$.

## A.2   DERIVATION OF EQUATION (2.29)

Referring to Figure 2.6, suppose $V_0^+$ is the forward voltage at the output port connecting to the load $Z_L$, with a reference direction pointing inside the TLS, and $V_0^-$ is the backward voltage at that port pointing outside the TLS. The scattering parameters $a_0$ and $b_0$ are then related to $V_0^+$ and $V_0^-$ as

$$
\begin{cases}
(a_0 + b_0)\sqrt{R_{n0}} = V_0^+ + V_0^- \\
(a_0 - b_0)/\sqrt{R_{n0}} = (V_0^+ - V_0^-)/R_{c1}
\end{cases}
\tag{A.8}
$$

$V_0^+$ in (A.8) can be eliminated by using (A.9) below, leading to (A.10):

$$
V_0^+ = \Gamma V_0^-, \text{ where } \Gamma = (Z_L - R_{c1})/(Z_L + R_{c1})
\tag{A.9}
$$

$$\begin{cases} (a_0 + b_0)\sqrt{R_{n0}} = (\Gamma + 1)V_0^- \\ (a_0 - b_0)/\sqrt{R_{n0}} = (\Gamma - 1)V_0^- / R_{c1} \end{cases} . \tag{A.10}$$

By eliminating $V_0^-$ in (A.10), and replacing $\Gamma$ by $(Z_L - R_{c1})/(Z_L + R_{c1})$, $a_0$ can be expressed in terms of $b_0$ as

$$a_0 = b_0(Z_L - R_{n0})/(Z_L + R_{n0}) . \tag{A.11}$$

## A.3   DERIVATION OF EQUATION (2.30)

Referring to Figure 2.6, the Kirchoff's second rule is applied to relate $a_N$ and $b_N$ through the equation

$$(a_N - b_N)R_s/\sqrt{R_{nN}} + (a_N + b_N)\sqrt{R_{nN}} = V_s \tag{A.12}$$

which may be solved for $a_N$ to obtain

$$a_N = \left[V_s\sqrt{R_{nN}} + (R_s - R_{nN})b_N\right]/(R_s + R_{nN}) . \tag{A.13}$$

# APPENDIX B

# Maxima Source Code

## B.1   MAXIMA SOURCE CODE FOR A PEC-SI

The source code file "deriveTEforPEC.max" listed below gives the analytical expressions the reflected and transmitted waves for a PEC-SI excited by a TE mode plane wave.

```
/*
*Maxima 5.12.0 batch file.
*Compute transmitted and reflected waves for a PEC strip interface.
*Excitation is TE mode plane wave.
*batch("deriveTEforPEC.max") to run.
*/

load("vect")$
writefile("resultTEforPEC.txt")$

/*unit vectors for the strip, vect_p is along the strip and vect_o othogonal to it.*/
vect_p: [cos(phs), sin(phs), 0]$
vect_o: express([0, 0, 1] ~ vect_p)$

/*unit vectors of propagation directions for incident, reflected, and transmitted waves*/
vect_ki: [−sin(thi)*cos(ph), −sin(thi)*sin(ph), −cos(thi)]$
vect_kr: [−sin(thi)*cos(ph), −sin(thi)*sin(ph), +cos(thi)]$
vect_kt: [−sin(tht)*cos(ph), −sin(tht)*sin(ph), −cos(tht)]$

/*TE mode for the incident wave, with an unit magnitude for electric field*/
vect_phi: [−sin(ph), cos(ph), 0]$
vect_Ei:  vect_phi$
vect_Er:  [Erx, Ery, Erz]$
vect_Et:  [Etx, Ety, Etz]$

vect_Hi: express(vect_ki ~ vect_Ei)/Z1$
vect_Hr: express(vect_kr ~ vect_Er)/Z1$
vect_Ht: express(vect_kt ~ vect_Et)/Z2$

Eq1: vect_p . (vect_Ei + vect_Er) = 0$
Eq2: vect_p . vect_Et = 0$
Eq3: vect_o . (vect_Ei + vect_Er − vect_Et) = 0$
Eq4: vect_p . (vect_Hi + vect_Hr − vect_Ht) = 0$
Eq5: vect_Er . vect_kr = 0$
Eq6: vect_Et . vect_kt = 0$
RESULT: linsolve([Eq1,Eq2,Eq3,Eq4,Eq5,Eq6], [Erx, Ery, Erz, Etx, Ety, Etz]), globalsolve:true$
Erx: trigsimp(Erx)$
Ery: trigsimp(Ery)$
Erz: trigsimp(Erz)$
Etx: trigsimp(Etx)$
Ety: trigsimp(Ety)$
Etz: trigsimp(Etz)$
```

```
/*obtain reflection and transmission coefficients for both TE and TM modes*/
Er:     [Erx, Ery, Erz]$
Er_TE: vect_phi . Er$
Er_TM: Er . [cos(ph), sin(ph), 0]$

Et:     [Etx, Ety, Etz]$
Et_TE: vect_phi . Et$
Et_TM: Et . [cos(ph), sin(ph), 0]$

/*simplify the results*/
Er_TE: trigreduce(trigreduce(trigsimp(Er_TE), ph), phs)$
Er_TM: trigreduce(trigreduce(trigsimp(Er_TM), ph), phs)$
Et_TE: trigreduce(trigreduce(trigsimp(Et_TE), ph), phs)$
Et_TM: trigreduce(trigreduce(trigsimp(Et_TM), ph), phs)$
Er_TE: scsimp(Er_TE, cos(2*ph)*cos(2*phs)+sin(2*ph)*sin(2*phs)=cos(2*ph−2*phs),
        cos(2*ph)*sin(2*phs)−sin(2*ph)*cos(2*phs)=sin(2*phs−2*ph))$
Er_TM: scsimp(Er_TM, cos(2*ph)*cos(2*phs)+sin(2*ph)*sin(2*phs)=cos(2*ph−2*phs),
        cos(2*ph)*sin(2*phs)−sin(2*ph)*cos(2*phs)=sin(2*phs−2*ph))$
Et_TE: scsimp(Et_TE, cos(2*ph)*cos(2*phs)+sin(2*ph)*sin(2*phs)=cos(2*ph−2*phs),
        cos(2*ph)*sin(2*phs)−sin(2*ph)*cos(2*phs)=sin(2*phs−2*ph))$
Et_TM: scsimp(Et_TM, cos(2*ph)*cos(2*phs)+sin(2*ph)*sin(2*phs)=cos(2*ph−2*phs),
        cos(2*ph)*sin(2*phs)−sin(2*ph)*cos(2*phs)=sin(2*phs−2*ph))$

Er_TE: rat(Er_TE);
Er_TM: rat(Er_TM);
Et_TE: rat(Et_TE);
Et_TM: rat(Et_TM);

closefile()$
```

The source code file "deriveTMforPEC.max" listed below gives the analytical expressions the reflected and transmitted waves for a PEC-SI excited by a TM mode plane wave.

```
/*
*Maxima 5.12.0 batch file.
*Compute transmitted and reflected waves for a PEC strip interface.
*Excitation is TM mode plane wave.
*batch("deriveTMforPEC.max") to run.
*/

load("vect")$
writefile("resultTMforPEC.txt")$

/*unit vectors for the strip, vect_p is along the strip and vect_o othogonal to it.*/
vect_p: [cos(phs), sin(phs), 0]$
vect_o: express([0, 0, 1] ~ vect_p)$

/*unit vectors of propagation directions for incident, reflected, and transmitted waves*/
vect_ki: [−sin(thi)*cos(ph), −sin(thi)*sin(ph), −cos(thi)]$
vect_kr: [−sin(thi)*cos(ph), −sin(thi)*sin(ph), +cos(thi)]$
vect_kt: [−sin(tht)*cos(ph), −sin(tht)*sin(ph), −cos(tht)]$

/*TM mode for the incident wave, with an unit magnitude for magnetic field*/
```

```
vect_phi: [−sin(ph), cos(ph), 0]$
vect_Hi: −vect_phi$
vect_Hr: [Hrx, Hry, Hrz]$
vect_Ht: [Htx, Hty, Htz]$


vect_Ei: express(vect_Hi ~ vect_ki)*Z1$
vect_Er: express(vect_Hr ~ vect_kr)*Z1$
vect_Et: express(vect_Ht ~ vect_kt)*Z2$


Eq1: vect_p . (vect_Ei + vect_Er) = 0$
Eq2: vect_p . vect_Et = 0$
Eq3: vect_o . (vect_Ei + vect_Er − vect_Et) = 0$
Eq4: vect_p . (vect_Hi + vect_Hr − vect_Ht) = 0$
Eq5: vect_Hr . vect_kr = 0$
Eq6: vect_Ht . vect_kt = 0$
RESULT: linsolve([Eq1,Eq2,Eq3,Eq4,Eq5,Eq6], [Hrx, Hry, Hrz, Htx, Hty, Htz]), globalsolve:true$
Hrx: trigsimp(Hrx)$
Hry: trigsimp(Hry)$
Hrz: trigsimp(Hrz)$
Htx: trigsimp(Htx)$
Hty: trigsimp(Hty)$
Htz: trigsimp(Htz)$


/*obtain reflection and transmission coefficients for both TE and TM modes*/
Ei_TM: vect_Ei . [cos(ph), sin(ph), 0]$


Hr:    [Hrx, Hry, Hrz]$
Er:    express(Hr ~ vect_kr)*Z1$
Er_TM: Er . [cos(ph), sin(ph), 0]/Ei_TM$
Er_TE: Er . vect_phi/Ei_TM$


Ht:    [Htx, Hty, Htz]$
Et:    express(Ht ~ vect_kt)*Z2$
Et_TM: Et . [cos(ph), sin(ph), 0]/Ei_TM$
Et_TE: Et . vect_phi/Ei_TM$


/*simplify the results*/
Er_TE: trigreduce(trigreduce(trigsimp(Er_TE), ph), phs)$
Er_TM: trigreduce(trigreduce(trigsimp(Er_TM), ph), phs)$
Et_TE: trigreduce(trigreduce(trigsimp(Et_TE), ph), phs)$
Et_TM: trigreduce(trigreduce(trigsimp(Et_TM), ph), phs)$
Er_TE: scsimp(Er_TE, cos(2*ph)*cos(2*phs)+sin(2*ph)*sin(2*phs)=cos(2*ph−2*phs),
       cos(2*ph)*sin(2*phs)−sin(2*ph)*cos(2*phs)=sin(2*phs−2*phs))$
Er_TM: scsimp(Er_TM, cos(2*ph)*cos(2*phs)+sin(2*ph)*sin(2*phs)=cos(2*ph−2*phs),
       cos(2*ph)*sin(2*phs)−sin(2*ph)*cos(2*phs)=sin(2*phs−2*phs))$
Et_TE: scsimp(Et_TE, cos(2*ph)*cos(2*phs)+sin(2*ph)*sin(2*phs)=cos(2*ph−2*phs),
       cos(2*ph)*sin(2*phs)−sin(2*ph)*cos(2*phs)=sin(2*phs−2*phs))$
Et_TM: scsimp(Et_TM, cos(2*ph)*cos(2*phs)+sin(2*ph)*sin(2*phs)=cos(2*ph−2*phs),
       cos(2*ph)*sin(2*phs)−sin(2*ph)*cos(2*phs)=sin(2*phs−2*phs))$


Er_TE: rat(Er_TE);
Er_TM: rat(Er_TM);
Et_TE: rat(Et_TE);
Et_TM: rat(Et_TM);


closefile()$
```

## B.2    MAXIMA SOURCE CODE FOR A PMC-SI

The source code file "deriveTEforPMC.max" listed below gives the analytical expressions the reflected and transmitted waves for a PMC-SI excited by a TE mode plane wave.

```
/*
*Maxima 5.12.0 batch file.
*Compute transmitted and reflected waves for a PMC strip interface.
*Excitation is TE mode plane wave.
*batch("deriveTEforPMC.max") to run
*/

load("vect")$
writefile("resultTEforPMC.txt")$

/*unit vectors for the strip, vect_p is along the strip and vect_o othogonal to it.*/
vect_p: [cos(phs), sin(phs), 0]$
vect_o: express([0, 0, 1] ~ vect_p)$

/*unit vectors of propagation directions for incident, reflected, and transmitted waves*/
vect_ki: [−sin(th2)*cos(ph), −sin(th2)*sin(ph), −cos(th2)]$
vect_kr: [−sin(th2)*cos(ph), −sin(th2)*sin(ph), +cos(th2)]$
vect_kt: [−sin(th1)*cos(ph), −sin(th1)*sin(ph), −cos(th1)]$

/*TE mode for the incident wave, with an unit magnitude for electric field*/
vect_phi: [−sin(ph), cos(ph), 0]$
vect_Ei:   vect_phi$
vect_Er:   [Erx, Ery, Erz]$
vect_Et:   [Etx, Ety, Etz]$

vect_Hi: express(vect_ki ~ vect_Ei)/Z2$
vect_Hr: express(vect_kr ~ vect_Er)/Z2$
vect_Ht: express(vect_kt ~ vect_Et)/Z1$

Eq1: vect_p . (vect_Hi + vect_Hr) = 0$
Eq2: vect_p . vect_Ht = 0$
Eq3: vect_o . (vect_Hi + vect_Hr − vect_Ht) = 0$
Eq4: vect_p . (vect_Ei + vect_Er − vect_Et) = 0$
Eq5: vect_Er . vect_kr = 0$
Eq6: vect_Et . vect_kt = 0$
RESULT: linsolve([Eq1,Eq2,Eq3,Eq4,Eq5,Eq6], [Erx, Ery, Erz, Etx, Ety, Etz]), globalsolve:true$
Erx: trigsimp(Erx)$
Ery: trigsimp(Ery)$
Erz: trigsimp(Erz)$
Etx: trigsimp(Etx)$
Ety: trigsimp(Ety)$
Etz: trigsimp(Etz)$

/*obtain reflection and transmission coefficients for both TE and TM modes*/
Er:     [Erx, Ery, Erz]$
Er_TE: vect_phi . Er$
```

```
Er_TM: Er . [cos(ph), sin(ph), 0]$

Et:     [Etx, Ety, Etz]$
Et_TE: vect_phi . Et$
Et_TM: Et . [cos(ph), sin(ph), 0]$

/*simplify the results*/
Er_TE: trigreduce(trigreduce(trigsimp(Er_TE), ph), phs)$
Er_TM: trigreduce(trigreduce(trigsimp(Er_TM), ph), phs)$
Et_TE: trigreduce(trigreduce(trigsimp(Et_TE), ph), phs)$
Et_TM: trigreduce(trigreduce(trigsimp(Et_TM), ph), phs)$
Er_TE: scsimp(Er_TE, cos(2*ph)*cos(2*phs)+sin(2*ph)*sin(2*phs)=cos(2*ph-2*phs),
        cos(2*ph)*sin(2*phs)-sin(2*ph)*cos(2*phs)=sin(2*phs-2*ph))$
Er_TM: scsimp(Er_TM, cos(2*ph)*cos(2*phs)+sin(2*ph)*sin(2*phs)=cos(2*ph-2*phs),
        cos(2*ph)*sin(2*phs)-sin(2*ph)*cos(2*phs)=sin(2*phs-2*ph))$
Et_TE: scsimp(Et_TE, cos(2*ph)*cos(2*phs)+sin(2*ph)*sin(2*phs)=cos(2*ph-2*phs),
        cos(2*ph)*sin(2*phs)-sin(2*ph)*cos(2*phs)=sin(2*phs-2*ph))$
Et_TM: scsimp(Et_TM, cos(2*ph)*cos(2*phs)+sin(2*ph)*sin(2*phs)=cos(2*ph-2*phs),
        cos(2*ph)*sin(2*phs)-sin(2*ph)*cos(2*phs)=sin(2*phs-2*ph))$

Er_TE: rat(Er_TE);
Er_TM: rat(Er_TM);
Et_TE: rat(Et_TE);
Et_TM: rat(Et_TM);

closefile()$
```

The source code file "deriveTMforPMC.max" listed below gives the analytical expressions the reflected and transmitted waves for a PMC-SI excited by a TM mode plane wave.

```
/*
*Maxima 5.12.0 batch file.
*Compute transmitted and reflected waves for a PMC strip interface.
*Excitation is TM mode plane wave.
*batch("deriveTMforPMC.max") to run
*/

load("vect")$
writefile("resultTMforPMC.txt")$

/*unit vectors for the strip, vect_p is along the strip and vect_o othogonal to it.*/
vect_p: [cos(phs), sin(phs), 0]$
vect_o: express([0, 0, 1] ~ vect_p)$

/*unit vectors of propagation directions for incident, reflected, and transmitted plane waves*/
vect_ki: [−sin(th2)*cos(ph), −sin(th2)*sin(ph), −cos(th2)]$
vect_kr: [−sin(th2)*cos(ph), −sin(th2)*sin(ph), +cos(th2)]$
vect_kt: [−sin(th1)*cos(ph), −sin(th1)*sin(ph), −cos(th1)]$

/*TM mode for the incident wave, with an unit magnitude for magnetic field*/
vect_phi: [−sin(ph), cos(ph), 0]$
vect_Hi: −vect_phi$
vect_Hr: [Hrx, Hry, Hrz]$
vect_Ht: [Htx, Hty, Htz]$
```

```
vect_Ei: express(vect_Hi ~ vect_ki)*Z2$
vect_Er: express(vect_Hr ~ vect_kr)*Z2$
vect_Et: express(vect_Ht ~ vect_kt)*Z1$

Eq1: vect_p . (vect_Hi + vect_Hr) = 0$
Eq2: vect_p . vect_Ht = 0$
Eq3: vect_o . (vect_Hi + vect_Hr − vect_Ht) = 0$
Eq4: vect_p . (vect_Ei + vect_Er − vect_Et) = 0$
Eq5: vect_Hr . vect_kr = 0$
Eq6: vect_Ht . vect_kt = 0$
RESULT: linsolve([Eq1,Eq2,Eq3,Eq4,Eq5,Eq6], [Hrx, Hry, Hrz, Htx, Hty, Htz]), globalsolve:true$
Hrx: trigsimp(Hrx)$
Hry: trigsimp(Hry)$
Hrz: trigsimp(Hrz)$
Htx: trigsimp(Htx)$
Hty: trigsimp(Hty)$
Htz: trigsimp(Htz)$

/*obtain reflection and transmission coefficients for both TE and TM modes*/
Ei_TM: vect_Ei . [cos(ph), sin(ph), 0]$

Hr: [Hrx, Hry, Hrz]$
Er: express(Hr ~ vect_kr)*Z2$
Er_TM: Er.[cos(ph), sin(ph), 0]/Ei_TM$
Er_TE: Er.vect_phi/Ei_TM$

Ht:    [Htx, Hty, Htz]$
Et:    express(Ht ~ vect_kt)*Z1$
Et_TM: Et.[cos(ph), sin(ph), 0]/Ei_TM$
Et_TE: Et.vect_phi/Ei_TM$

/*simplify the results*/
Er_TE: trigreduce(trigreduce(trigsimp(Er_TE), ph), phs)$
Er_TM: trigreduce(trigreduce(trigsimp(Er_TM), ph), phs)$
Et_TE: trigreduce(trigreduce(trigsimp(Et_TE), ph), phs)$
Et_TM: trigreduce(trigreduce(trigsimp(Et_TM), ph), phs)$
Er_TE: scsimp(Er_TE, cos(2*ph)*cos(2*phs)+sin(2*ph)*sin(2*phs)=cos(2*ph−2*phs),
        cos(2*ph)*sin(2*phs)−sin(2*ph)*cos(2*phs)=sin(2*phs−2*ph))$
Er_TM: scsimp(Er_TM, cos(2*ph)*cos(2*phs)+sin(2*ph)*sin(2*phs)=cos(2*ph−2*phs),
        cos(2*ph)*sin(2*phs)−sin(2*ph)*cos(2*phs)=sin(2*phs−2*ph))$
Et_TE: scsimp(Et_TE, cos(2*ph)*cos(2*phs)+sin(2*ph)*sin(2*phs)=cos(2*ph−2*phs),
        cos(2*ph)*sin(2*phs)−sin(2*ph)*cos(2*phs)=sin(2*phs−2*ph))$
Et_TM: scsimp(Et_TM, cos(2*ph)*cos(2*phs)+sin(2*ph)*sin(2*phs)=cos(2*ph−2*phs),
        cos(2*ph)*sin(2*phs)−sin(2*ph)*cos(2*phs)=sin(2*phs−2*ph))$

Er_TE: rat(Er_TE);
Er_TM: rat(Er_TM);
Et_TE: rat(Et_TE);
Et_TM: rat(Et_TM);

closefile()$
```

# Bibliography

[1] P. D. Maagt, R. Gonzalo, Y. C. Vardaxoglou, and J.-M. Baracco, "Electromagnetic Bandgap Antennas and Components for Microwave and (Sub)Millimeter Wave Applications," *IEEE Trans. Antennas Propagat.*, vol. 51, pp. 2667–2677, Oct. 2003. DOI: 10.1109/TAP.2003.817566

[2] M. Thèvenot, C. Cheype, A. Reineix, and B. Jecko, "Directive Photonic-Bandgap Antennas," *IEEE Trans. Microw. Theory Tech.*, vol. 47, pp. 2115–2122, Nov. 1999. DOI: 10.1109/22.798007

[3] C. Serier, C. Cheype, R. Chantalat, M. Thèvenot, T. Monédière, A. Reineix, and B. Jecko, "1-D Photonic Bandgap Resonator Antenna," *Microwave Opt. Technol. Lett.*, vol. 29, no. 5, pp. 312–315, June 5, 2001. DOI: 10.1002/mop.1164

[4] T. Akalin, J. Danglot, O. Vanbésien, and D. Lippens, "A Highly Directive Dipole Antenna Embedded in a Fabry-Pérot Type Cavity," *IEEE Microw. Wireless Compon. Lett.*, vol. 12, pp. 48–50, Feb. 2002. DOI: 10.1109/7260.982873

[5] R. Gardelli, M. Albani, and F. Capolino, "EBG Superstrates for Dual Polarized Sparse Arrays," *Proc. IEEE AP-S Int. Symp.*, Washington, D.C., vol. 2A, pp. 586–589, July 3-8, 2005. DOI: 10.1109/APS.2005.1551878

[6] A. R. Weily, K. P. Esselle, B. C. Sanders, and T. S. Bird, "Circularly Polarized 1-D EBG Resonator Antenna," *Proc. 10th Int. Symp. Antenna Technol. Appl. Electromagn. URSI Conf.*, Fairmont Château Laurier, Ottawa, ON, Canada, pp. 405–408, July 20-23, 2004.

[7] H. Y. Yang and N. G. Alexopoulos, "Gain Enhancement Methods for Printed Circuit Antennas Through Multiple Superstrates," *IEEE Trans. Antennas Propagat.*, vol. AP-35, pp. 860–863, July 1987.

[8] X. H. Wu, A. A. Kishk, and A. W. Glisson, "A Transmission Line Method to Compute the Far-Field Radiation of Arbitrarily Directed Hertzian Dipoles in a Multilayer Dielectric Structure: Theory and Applications," *IEEE Trans. Antennas Propagat.*, vol. 54, pp. 2731–2741, Oct. 2006.

[9] X. H. Wu, A. A. Kishk, and A. W. Glisson, "A Transmission Line Method to Compute the Far-Field Radiation of Arbitrary Hertzian Dipoles in a Multilayer Structure Embedded With PEC Strip Interfaces," *IEEE Trans. Antennas Propagat.*, vol. 55, pp. 3191–3198, Nov. 2007. DOI: 10.1109/TAP.2007.9088368

[10] "IE3D, 11.0," Zeland Software Inc., Fremont, CA.

[11] "HFSS, 10.0," Ansoft Corporate, Pittsburgh, PA.

[12] T. S. Sijher and A. A. Kishk, "Antenna Modeling by Infinitesimal Dipoles Using Genetic Algorithms," *Progr. Electromagn. Res.*, PIER 52, pp. 225–254, 2005. DOI: 10.2528/PIER04081801

[13] "WIPL-D, Pro v4.1," WIPL-D Ltd., Belgrade, Serbia.

[14] J. Robinson and Y. Rahmat-Samii, "Particle Swarm Optimization in Electromagnetics," *IEEE Trans. Antennas Propagat.*, vol. 52, pp. 397–407, Feb. 2004. DOI: 10.1109/TAP.2004.823969

[15] S. M. Mikki and A. A. Kishk, "Investigation of the Quantum Particle Swarm Optimization Technique for Electromagnetic Applications," *Proc. IEEE AP-S Int. Symp.*, Washington, D.C., vol. 2A, pp. 45-48, July 3-8, 2005, DOI: doi:10.1109/APS.2005.1551731

[16] A. A. Kishk, "Analysis of Hard Surfaces of Cylindrical Structures of Arbitrarily Shaped Cross Section Using Asymptotic Boundary Conditions," *IEEE Trans. Antennas Propagat.*, vol. 51, pp. 1150–1156, June 2003. DOI: 10.1109/TAP.2003.812270

[17] P.-S. Kildal, "Artificially Soft and Hard Surfaces in Electromagnetics," *IEEE Trans. Antennas Propagat.*, vol. 38, pp. 1537–1544, Oct. 1990. DOI: 10.1109/8.59765

[18] A. A. Kishk and P.-S. Kildal, "Asymptotic Boundary Conditions for Strip-loaded Scatterers Applied to Circular Dielectric Cylinders Under Oblique Incidence," *IEEE Trans. Antennas Propagat.*, vol. 45, pp. 51–56, Jan. 1997. DOI: 10.1109/8.554240

[19] "Maxima, a computer algebra system," http://maxima.sourceforge.net/

[20] A. A. Kishk and L. Shafai, "Gain Enhancement of Antennas over Finite Ground Plane Covered by a Dielectric Sheet," *IEE Proc., Part H*, vol. 134, pp. 60–64, Feb. 1987.

[21] A. A. Kishk, "One-dimensional Electromagnetic Bandgap for Directivity Enhancement of Waveguide Antennas," *Microw. Opt. Technol. Lett.*, vol. 47, no. 5, pp. 430–434, 2005. DOI: 10.1002/mop.21192

[22] S. M. Mikki and A. A. Kishk, "Theory and Applications of Infinitesimal Dipole Models for Computational Electromagnetics," *IEEE Trans. Antennas Propagat.*, vol. 55, pp. 1325–1337, May 2007. DOI: 10.1109/TAP.2007.895625

[23] X. H. Wu, A. A. Kishk, and A. W. Glisson, "Modeling of Wideband Antennas by Frequency-Dependent Hertzian Dipoles," *IEEE Trans. Antennas Propagat.*, vol. 56, pp. 2481–2489, Aug. 2008. DOI: 10.1109/TAP.2008.927546

[24] C. A. Balanis, *Advanced Engineering Electromagnetics*, John Wiley & Sons, 1989.

[25] "FEKO, 5.2" Stellenbosch, South Africa.

# Biography

**Xuan Hui Wu** received a Bachelor's in Engineering degree from the Department of Information Science & Electronic Engineering (ISEE), Zhejiang University, China, in 2001, a Masters in Engineering degree from the Department of Electrical and Computer Engineering (ECE), National University of Singapore, Singapore, in 2005, and a Ph.D. degree from the Department of Electrical Engineering, University of Mississippi.

In July 2002, he joined the Institute for Infocomm Research ($I^2R$), Agency of Science Technology and Research, Singapore, as a Research Graduate Student. Since 2004, he is with the Microwave Research Laboratory, University of Mississippi. His current research interests include computational electromagnetics, optimizations in electromagnetics, ultrawide-band radio systems, and multiple-input-multiple-output systems. Dr. Wu is a member of Sigma Xi Society and a member of Phi Kappa Phi Society.

**Ahmed A. Kishk** is a Professor of Electrical Engineering, University of Mississippi (since 1995). He was an Associate Editor of *Antennas and Propagation Magazine* from 1990–1993, and is now the Editor. He was a co-editor of the special issue on "Advances in the Application of the Method of Moments to Electromagnetic Scattering Problems" in the *ACES Journal*. He was also an editor of the ACES Journal during 1997. He was an Editor-in-Chief of the ACES Journal from 1998–2001. He was the chair of Physics and Engineering division of the Mississippi Academy of Science (2001–2002). He was a guest Editor of the special issue on artificial magnetic conductors, soft/hard surfaces, and other complex surfaces, on the *IEEE Transactions on Antennas and Propagation*, January 2005.

His research interest includes the areas of design of millimeter frequency antennas, feeds for parabolic reflectors, dielectric resonator antennas, microstrip antennas, soft and hard surfaces, phased array antennas, and computer aided design for antennas. He has published over 150 refereed journal articles and book chapters. He is a co-author of the *Microwave Horns and Feeds* book (London, IEE, 1994; New York: IEEE, 1994) and a co-author of Chapter 2 in *Handbook of Microstrip Antennas* (Peter Peregrinus Limited, United Kingdom, J. R. James and P. S. Hall, Ed., Ch. 2, 1989). Dr. Kishk received the 1995 and 2006 outstanding paper award for papers published in the *Applied Computational Electromagnetic Society Journal*. He received the 1997 Outstanding Engineering Educator Award from Memphis section of the IEEE. He received the Outstanding Engineering Faculty Member of the 1998 and the Award of Distinguished Technical Communication for the entry of IEEE Antennas and Propagation Magazine, 2001. He received the 2001 and 2005 Faculty research award for outstanding performance in research. He also received The Valued Contribution Award for outstanding Invited Presentation, "EM Modeling of Surfaces with STOP or GO Characteristics

- Artificial Magnetic Conductors and Soft and Hard Surfaces" from the Applied Computational Electromagnetic Society. He received the Microwave Theory and Techniques Society Microwave Prize 2004. Dr. Kishk is a Fellow member of IEEE since 1998 (Antennas and Propagation Society and Microwave Theory and Techniques), a member of Sigma Xi society, a member of the U.S. National Committee of International Union of Radio Science (URSI) Commission B, a member of the Applied Computational Electromagnetics Society, a member of the Electromagnetic Academy, and a member of Phi Kappa Phi Society.

**Allen W. Glisson** is a Professor and Chair of the Department of Electrical Engineering, University of Mississippi. He received B.S., M.S., and Ph.D. degrees in electrical engineering from the University of Mississippi, in 1973, 1975, and 1978, respectively. In 1978, he joined the faculty of the University of Mississippi. He was selected as the Outstanding Engineering Faculty Member in 1986, 1996, and 2004. He received a Ralph R. Teetor Educational Award in 1989 and in 2002 he received the Faculty Service Award in the School of Engineering. His current research interests include the development and application of numerical techniques for treating electromagnetic radiation and scattering problems, and modeling of dielectric resonators and dielectric resonator antennas.

Dr. Glisson is a Fellow of the IEEE, a Fellow of the Applied Computational Electromagnetics Society, and a member of Commission B of the International Union of Radio Science. He has received a best paper award from the SUMMA Foundation and twice received a citation for excellence in refereeing from the American Geophysical Union. He was a recipient of the 2004 Microwave Prize awarded by the Microwave Theory and Techniques Society and received the 2006 Best Paper Award from the *Applied Computational Electromagnetics Society Journal*. He served as the Associate Editor for Book Reviews and Abstracts for the *IEEE Antennas and Propagation Society Magazine* from 1984 until 2006. He has served on the Board of Directors of the Applied Computational Electromagnetics Society, is currently Treasurer of ACES, and is a member of the AP-S IEEE Press Liaison Committee. He has previously served as a member of the IEEE Antennas and Propagation Society Administrative Committee, as the secretary of Commission B of the U.S. National Committee of URSI, as an Associate Editor for *Radio Science*, as Co-Editor-in-Chief of the *Applied Computational Electromagnetics Society Journal*, and as the Editor-in-Chief of the *IEEE Transactions on Antennas and Propagation*.

Printed in the United States
by Baker & Taylor Publisher Services